Misconceptions in Chemistry

Hans-Dieter Barke · Al Hazari · Sileshi Yitbarek

Misconceptions in Chemistry

Addressing Perceptions in Chemical Education

Prof. Dr. Hans-Dieter Barke
Universität Münster
FB 12 Chemie und Pharmazie
Inst. Didaktik der Chemie
Fliednerstr. 21
48149 Münster
Germany

Dr. Al Hazari
University of Tennessee
Dept. Chemistry
505 Buehler Hall
Knoxville TN 37996-1600
USA

Dr. Sileshi Yitbarek
Kotebe College of Teacher
Education
Addis Ababa
Ethiopia

ISBN: 978-3-540-70988-6

e-ISBN: 978-3-540-70989-3

DOI 10.1007/978-3-540-70989-3

Library of Congress Control Number: 2008933285

© Springer-Verlag Berlin Heidelberg 2009

This work is subject to copyright. All rights are reserved, whether the whole or part of the material is concerned, specifically the rights of translation, reprinting, reuse of illustrations, recitation, broadcasting, reproduction on microfilm or in any other way, and storage in data banks. Duplication of this publication or parts thereof is permitted only under the provisions of the German Copyright Law of September 9, 1965, in its current version, and permission for use must always be obtained from Springer. Violations are liable to prosecution under the German Copyright Law.

The use of general descriptive names, registered names, trademarks, etc. in this publication does not imply, even in the absence of a specific statement, that such names are exempt from the relevant protective laws and regulations and therefore free for general use.

Cover design: KünkelLopka GmbH

Printed on acid-free paper

9 8 7 6 5 4 3 2 1

springer.com

We would like to dedicate this book to Chemistry students everywhere. May their quest to full understanding of this subject lead them to the discovery of the truth and the beauty of Chemistry.

Acknowledgment

First, we would like to cordially thank **Dr. Hilde Wirbs**, wife of Hans-Dieter Barke, for all the discussions concerning chemistry misconceptions and for testing some proposed strategies for teaching and learning in her school, Kaethe-Kollwitz Realschule in Emsdetten near Muenster, Germany. Many thanks to colleagues and friends: **Prof. Dr. Dieter Sauermann** (Munich), **Prof. Dr. Guenther Harsch** (Muenster) and **Prof. Dr. Rebekka Heimann** (Leipzig) who helped us get interesting insights into problems of chemistry education.

Special appreciation goes to **Barbara Doran–Rogel** and **Dr. Friedhelm Rogel** for translating all chapters of the German book on "Chemical Misconceptions: Prevention and Cure", written by Hans-Dieter Barke into English. Same appreciation belongs to artist Mr. Mulugeta Gebrekidan, he did the drawings with regard to the concept cartoons.

We are also gratefully indebted to Dr. Temechegn Engida, Dr. Birte Moeller, Dr. Nina Strehle, Dr. Cosima Kuhl, Dr. Wahyne Sopandi, Dr. Sebastian Musli, Dr. Claus Hilbing, Reinhard Roelleke, Tobias Doerfler and Serkalem Girma for availing themselves whenever we needed their support.

Finally we would like to thank the **Fonds of Chemical Industries FCI** in Frankfurt, Germany, and the **German Society of Research DFG** in Bonn, Germany. The financial support of the FCI allowed us to realize and to utilize all the Masters theses as the research basis of the German book on Chemical Misconceptions. The DFG supported our visits to Chemical Education congresses in the United States of America where we met Dr. Al Hazari, University of Tennessee, Knoxville.

October 2008   Hans-Dieter Barke, Al Hazari, Sileshi

Contents

Introduction 1

1 Perceptions of Ancient Scientists 9
   1.1 The Theory of Basic Matter 10
   1.2 Transformation Concepts of the Alchemists 10
   1.3 The Phlogiston Theory 11
   1.4 Historic Acid–Base Theories 12
   1.5 "Horror Vacui" and the Particle Concept 14
   1.6 Atoms and the Structure of Matter 15
   References 20

2 Students' Misconceptions and How to Overcome Them 21
   2.1 Students' Preconcepts 21
   2.2 School-Made Misconceptions 24
   2.3 Students' Concepts and Scientific Language 26
   2.4 Effective Strategies for Teaching and Learning 28
   References 33
   Further Reading 34

3 Substances and Properties 37
   3.1 Animistic Modes of Speech 38
   3.2 Concepts of Transformation 39
   3.3 Concepts of Miscibility for Compounds 41
   3.4 Concepts of Destruction 43
   3.5 Concepts of Combustion 46
   3.6 Concepts of "Gases as not Substances" 50
   3.7 Experiments on Substances and Their Properties 52
   References 64
   Further Reading 65

4 Particle Concept of Matter 67
   4.1 Smallest Particles of Matter and Mental Models 69
   4.2 Preformed and Non-preformed Particles 73

	4.3	Smallest Particles as Portions of Matter	76
	4.4	Particles and the "Horror Vacui"	78
	4.5	Particles – Generic Term for Atoms, Ions and Molecules	82
	4.6	Formation of Particles and Spatial Ability	83
	4.7	Diagnosis Test for Understanding the Particle Model of Matter	86
	4.8	Experiments on Particle Model of Matter	93
		References	99
		Further Reading	100
5	**Structure–Property Relationships**		103
	5.1	Structure and Properties of Metals and Alloys	103
	5.2	Existence of Ions and Structure of Salts	108
	5.3	Mental Models on Ionic Bonding	115
	5.4	Chemical Structures and Symbolic Language	125
	5.5	Experiments on Structure–Property Relationships	130
		References	140
		Further Reading	142
6	**Chemical Equilibrium**		145
	6.1	Overview of the Most Common Misconceptions	145
	6.2	Empirical Research	146
	6.3	Teaching and Learning Suggestions	156
	6.4	Experiments on Chemical Equilibrium	165
		References	169
		Further Reading	170
7	**Acid–Base Reactions**		173
	7.1	Acid–Base Reactions and the Proton Transfer	173
	7.2	Misconceptions	175
	7.3	Teaching and Learning Suggestions	183
	7.4	Experiments on Acids and Bases	193
		References	204
		Further Reading	204
8	**Redox Reactions**		207
	8.1	Misconceptions	209
	8.2	Teaching and Learning Suggestions	217
	8.3	Experiments on Redox Reactions	226
		References	231
		Further Reading	232
9	**Complex Reactions**		235
	9.1	Misconceptions	237
	9.2	Teaching and Learning Suggestions	245

9.3	Experiments on Complex Reactions	252
	References	259

10 Energy 261
    10.1 Misconceptions 262
    10.2 Empirical Research 265
    10.3 Energy and Temperature 269
    10.4 Fuel and Chemical Energy 272
    10.5 Experiments on Energy 279
    References 286
    Further Reading 287

List of Experiments 289

Epilogue 293

Introduction

"At last I found a lecture worth getting up early in the morning for; excellent examples and experiments of teaching chemistry; now I know what chemistry education means and why it is so important for my studies; good to have the clear concept of the 'pie-chart' at the beginning of all lectures" [1]. These comments from would-be-chemistry teachers show that the lectures of chemistry education in our Institute of Chemistry Education at University of Muenster are extremely beneficial in assisting them in their approach to teaching chemistry at school.

The most important subjects of 15 lectures in chemistry education can be presented in a kind of "pie-chart" (see Fig. 0.1): "Learners ideas and misconceptions; experiments; structural and mental models; terminology, symbols and formulae; every-day-life chemistry; media; motivation; teaching aims" [1]. Because we want to put a lot of emphasis on the learner, she or he is therefore placed at the centre of the diagram. Secondly, "scientific ideas" should be reflected in association with appropriate "teaching processes" for the learner. Finally there should be reflections on the "human element" or "context" to each subject as Mahaffy [2] has proposed. There are free sectors in that diagram – for more chemistry education subjects to reflect upon. In this book emphasis is given to students' preconceptions and misconceptions; experiments; structural and mental models; terminology, symbols and formulae.

In our experience of beginning of courses in chemistry education, would-be-chemistry teachers are often not clear about their own or students' "Preconcepts" or "misconceptions". They are not aware of how important it is to know more about these concepts and how to integrate them into chemistry education at school. Our reason for publishing this book is to assist those studying to become chemistry teachers and those already teaching chemistry at school. We also support professor Jung's comments, a physics teacher in Germany: "One should really write a book on diagnosing misconceptions and give it to all teachers". The psychologist Langthaler made similar comments: "If you, as a teacher, would have more diagnostic abilities and tools, many problems with your students would never even arise".

In planning coursework in the past few decades, teachers were under the impression that young pupils had hardly any knowledge of science. Therefore,

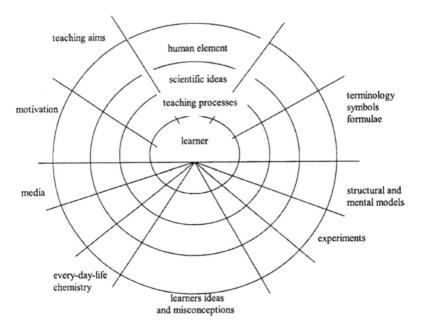

Fig. 0.1 Main subjects of a lecture in chemistry education, "pie-chart" metaphor [1]

teachers had only to decide how to plan a lecture in order to transmit scientific ideas to their pupils, perhaps incorporating laboratory experiments or new technology-based methods.

However, research has found otherwise. Latest studies in science education show that children and adolescents have many images and ideas about nature and their own surroundings. Nieswandt [3], for example, states according to her research: "Even from the very early days children are developing their own ideas about nature and every-day life. They are looking at cause and effect, at what happens if they let something fall to the floor, if they push, pull or throw things. By repeating these experiences, they develop a concept about the movement of things on the ground and in the air. Examination of these ideas and concepts shows that they are often in conflict to those that are typically accepted by the scientific community. Research in the field of pre-school knowledge has shown that these preconceptions of learning and comprehension tend to hinder the understanding of modern scientific concepts".

Research in students' conceptions in chemistry is based on the constructivist approach to learning, in which students construct their own cognitive structure. According to this approach of learning, learners generate their own meaning based on their background, attitudes, abilities, experiences etc, before, during and after instruction. Since students do construct or build their own concepts, their constructions differ from the one that the instructor holds and has tried to present.

These different concepts are variously described by different researchers as: misconceptions, alternative conceptions, naïve beliefs, erroneous ideas, multiple private versions of science, underlying sources of error, personal models of reality, spontaneous reasoning, developing conceptions, misunderstanding, mistakes, misinterpretation of facts, personal constructs and persistent pitfalls – to name just a few [4–10]. The authors will use the term "misconceptions" for the simple reason that researchers refer to it more often.

In order to promote successful learning or at least to simplify it, science educators should diagnose which preconceived images and explanations students hold. In this regard, Treagust [11] suggests using specific questionnaires to diagnose misconceptions of content and basic ideas: "By using a diagnostic test at the beginning or upon completion of a specific science topic, a science teacher can obtain clearer ideas about the nature of students' knowledge and misconceptions in the topic" [11].

In this book, the authors present diagnostic instruments from scientific literature and from their own empirical research. They also want to propose strategies for teaching and learning chemistry, which may help to cure or even to prevent students misconceptions. Most of our basic research is done in Germany and has been published in the German book "Chemiedidaktik – Diagnose und Korrektur von Schülervorstellungen" [12] by Springer, Heidelberg, in the year 2006.

This book includes the translation of the German book on misconceptions and includes a lot of new supplements. Both books will give teacher educators, chemistry teachers and would-be-chemistry teachers various examples of students' preconceptions and misconceptions, for consideration in their lessons. With this knowledge, teachers are better able to plan their own questionnaires and interviews in order to find out specific preconceptions and misconceptions of their students. Teachers become more aware of such misconceptions and are able to discuss them in their classrooms. Once the alternative conceptions of the students have been identified, the teacher has to decide how to deal with them:

- giving the scientific idea first and then discussing misconceptions, or
- go over students' misconceptions first, make them uncomfortable with their own ideas and instruct the scientific concept afterwards.

Examples and proposals on how to deal with misconceptions during lectures and how to include convincing experiments are described in all chapters.

Gabel [13] found out that many teachers are not familiar with or do not acknowledge the science education research regarding misconceptions. Therefore, they do not intend to incorporate them into their lecture plan: "Probably nine out of ten instructors are not aware of the research on student misconceptions, or do not utilize ways to counteract these misconceptions in their instruction". Gilbert et al. [14] calls upon all teachers not only to increase their awareness of the diagnostic methods available for finding misconceptions, but also to implement them in their lessons. They also suggested that teachers should be aware of these diagnostic tools during their teacher-training curriculum: "The pre-service and in-service education of prospective and experienced chemistry teachers can play a crucial role in bridging the gap between chemical education research and classroom practice" [14]. In this regard, they point out: "Increasing chemistry teachers' awareness of chemical education research, improving the use of chemical education research findings and involving chemistry teachers in chemical education research" [14].

The variety of subjects in chemical education automatically leads to a large assortment of misconceptions that are not possible to cover in one book. For this reason, in this book, an attempt will be made to address the most important concepts that arise in chemistry education: "We have examined the question extensively to see if it is possible to limit the relevant chemistry subject contents to superior central themes or scientific terms which are paramount for the comprehension of most chemical processes. We have discussed this issue with university professors along with experts in the field of teaching as well as with teachers at all levels. Because of this discussion, we have come up with the following consensus: an elementary understanding of basic principles of chemistry can be traced back to the command of a limited amount of principles – these principles are necessary for understanding chemistry. We call these principles, Basic Concepts" [15]. The following **basic concepts of chemistry** will be the foundation of this book: (a) Substances and Properties, (b) Particle Model of Matter, (c) Structure and Properties, (d) Chemical Equilibrium, (e) Donor–Acceptor Principle, and (f) Energy. The Donor–Acceptor Principle will be differentiated to Acid–Base Reactions, Redox Reactions and Complex Reactions.

Studying all the misconceptions, one would find parallels between the thinking of our students today and that of **ancient scientists of the last centuries**: As one would expect, all observations in our every-day life, which were carried out in a similar fashion throughout the ages, lead us to similar interpretations. In Chap. 1, some of the theories of the ancient scientists are described. These may help to show how misconceptions have been corrected in history. An example is the well-known **Phlogiston Theory**. Scientists of the 17th and 18th century assumed that all combustible substances contain some "invisible matter" and that upon burning, SOMETHING of the burning substances dissipates in the air. The German scientist Stahl called this "Phlogiston" and the Phlogiston Theory became highly accepted by scientists of these ages.

In Chap. 2 student's misconceptions and strategies for overcoming them will be discussed in the sense that young people have mostly observed their environment in a right and proper way, that they are not responsible for their "mistakes". An expansion on this idea will be presented that – besides the known **preconcepts** – there are also **school-made misconceptions** which arise in advanced science courses and which do not stem so much from the learners but rather from the teachers and the textbooks; or from the specific complexity of some subjects.

In Chaps. 3–10 preconcepts and misconceptions of the basic concepts and teaching and learning strategies will be reflected. In addition, since laboratory experiments and mental models are very convincing tools that could cure misconceptions in the cognitive structure of students, many experiments and structural models will be described following every chapter.

Children's preconcepts develop without a particle concept, hence the important "matter-particle concept" is divided in this book into two levels: "Substances and Properties" (see Chap. 3) and "Particle Model of Matter" (see Chap. 4).

Substances and Properties. In general, without having had a specific education in the particle concept, children tend to develop their original ideas based on properties of matter and their transformations. The well-known destruction concept concerning combustion is explained with the idea that "something is released into the air", or the properties of substances may change in their mind: "copper becomes green, silver becomes black". Along these lines, many explanations of other observations and phenomena will be discussed in Chap. 3.

Particle Model of Matter. "In chemistry, one goes with the premise that all matter is composed of submicroscopic particles, namely atoms or ions. They can appear isolated (atoms in noble gases), but mostly combined in groups of atoms or ions. They more or less form large aggregates with specific characteristics (e.g. metal crystals or salt crystals). The variety of matter is created by the many possible combinations and structures of a limited number of elements, of atoms and ions" [15]. With respect to this, misconceptions can only be school-made because one needs special lessons about the particle model of matter or Dalton's atomic concept before one develops misconceptions related to these concepts. These discussions will follow in Chap. 4.

Structure and Properties. "The properties of substances are directly dependent on the type of particles and on aggregates of particles, respectively chemical structures. Herein, it is more important to look at how atoms and ions are arranged in special structures than which kind of particles are involved" [15]. Chapter 5 will deal with this.

Chemical Equilibrium. "The exchange of matter and energy is basically possible in two directions at the same time, in forward and backward reactions. As a result, there exists under certain constant conditions a defined relationship between reactants and products which one describes as chemical equilibrium" [15]. In Chap. 6 misconceptions according to equilibria are reflected, teaching-learning suggestions are offered for every basic concept.

Donor–Acceptor Principle. "Particle aggregates, but also the atomic units of matter themselves can interact or react with each other and can thereby develop attraction and repulsion forces, and can transfer particles or energy. These particles or energies are transferred from one partner to the other" [15]. The transmitted particles or energies can be (a) protons, (b) electrons, (c) ligands. In this respect, one can see school-made misconceptions with regard to (a) acid–base reactions (Chap. 7), (b) redox reactions (Chap. 8), and (c) complex reactions (Chap. 9).

Energy. "Energy is stored in all substances; the amount of stored chemical energy is a very characteristic property. In chemical reactions in which matter is transformed, chemical energies are also changing – matter either releases or absorbs energy" [15]. Chapter 10 will discuss this.

In order to identify the observational goal and to know the intention, the **experiments** will have the heading "Problem" with additional comments in the first paragraph; the second paragraph shows "Material, Procedure and Observation". The explanations of the observations are not written: they may be very different depending on the grade or progress of the specific class or student group. Because of short descriptions of all experiments only an experienced teacher may conduct the experiments – the descriptions are not for students or beginners.

Every chapter related to the basic concepts starts with a **"Concept Cartoon"** [16]: four students explain a special phenomenon from their view and present the correct answer and some misconceptions of the involved subject [16]. These cartoons summarize the most important misconceptions for the reader; they are also helpful in class to start the discussion about students' ideas. The students may find their own ideas among the cartoons or will get to know which other ideas are possible. In every case they can discuss ideas and misconceptions by asking the question "what do you think?" – and may find the scientific answer, with the help of their teacher or on their own. It is even possible to use the cartoons for assessment [16].

We hope that many teachers in the area of chemistry or science will be inspired to reflect misconceptions and related strategies for chemistry instruction. The authors wish everyone lots of success in their own studies and in educational ways to cure or prevent misconceptions – through suggestions and proposals for revising the curriculum and taking new roads. Thank you for teaching "modern chemistry" to our children and our future adults!

References

1. Barke, H.-D., Harsch, G.: Chemiedidaktik Heute. Lernprozesse in Theorie und Praxis. Berlin, Heidelberg, New York 2001 (Springer)
2. Mahaffy, P: Moving Chemistry Education in to 3D: A tetrahedral metaphor for understanding chemistry. Journal of Chemical Education 83 (2006), 1
3. Nieswandt, M.: Von Alltagsvorstellungen zu wissenschaftlichen Konzepten: Lernwege von Schuelerinnen und Schuerlern im einfuehrenden Chemieunterricht. ZfDN 7 (2001), 33
4. Blosser, P.: Science misconceptions research and some implications for the teaching to elementary school students. ERIC/SMEAC Science Education (1987), 1
5. Elizabeth, H.: The students Laboratory and the Science Curriculum. Chatham 1990
6. Eylon, B., Linn, M.: Learning and instruction: An examination of four research perspectives in science education. Review of Educational Research. 58 (1988), 251
7. Fensham, P. et al. (Ed): The Content of Science: A Constructive Approach to its Learning and Teaching. The Falmer Press 1994
8. McGuigan, L., Schilling, M.: Children learning in science. In: Cross, E.A., Peet, G.: Teaching Science in the Primary School, Book One: A Practical Source Book of Teaching Strategies. Northcote House Publishers Ltd 1997, 24

References

9. Nakhleh, B., Mary, M.: Why some students don't learn chemistry. Journal of Chemical Education 69 (1992), 191
10. Wandersee, J.H. et al.: Research on alternative conceptions in science. In: Gabel, D.: Handbook of Research on Science Teaching and Learning. New York 1994
11. Treagust, D.F.: Development and use of diagnostic tests to evaluate students' misconceptions in science. International Journal of Science Education 10 (1988), 159
12. Barke, H.-D.: Chemiedidaktik – Diagnose und Korrektur von Schuelervorstellungen. Heidelberg 2006 (Springer).
13. Gabel, D.: Improving teaching and learning through chemistry education research: A look to the future. Journal of Chemical Education 76 (1999), 548
14. Gilbert, J.K., Justi, R., Driel, J.H., De Jong, O., Treagust, D.J.: Securing a future for chemical education. CERP 5 (2004), 5
15. Buender, W., Demuth, R., Parchmann, I.: Basiskonzepte – welche chemischen Konzepte sollen Schueler kennen und nutzen? PdN-ChiS 52 (2003), Heft 1, 2
16. Temechegn, E., Sileshi, Y.: Concept Cartoons as a Strategy in Learning, Teaching and Assessment Chemistry. Addis Ababa, Ethiopia, 2004

Chapter 1
Perceptions of Ancient Scientists

Students' conceptions of combustion ("SOMETHING is going up into the air") amazingly account for parts of the historic **Phlogiston Theory**. Through identical observations, parallels have been noted between the beliefs of today's youth and many of the ancient scientists. It makes sense to study the development of some historic theoretical themes and examine how they are deep-rooted in science:

- theory of basic matter by the Greek philosophers,
- transformation concepts of the alchemists,
- the Phlogiston theory,
- historic acid–base theories,
- "horror vacui" and particle concept,
- atoms and the structure of matter, etc.

One should perhaps consider and use historic concepts to analyze historical conceptual changes, develop today's concepts of education and compare with those changes of the past. Moreover, the historical changes may be included in the teaching–learning strategies and materials; the students should talk about and realize that "their problems are similar to those of scientists of the past" [1]. Due to this, students would be more likely to let go of their own misconceptions: "If students are made aware of the misconceptions of earlier scientists, perhaps they might find their own misconceptions among them. If the teacher compares and contrasts the historical misconceptions with the current explanation, students may be convinced to discard their limited or inappropriate propositions and replace them with modern scientific ones" [2].

The above-mentioned themes can be thought through historical problem-oriented approach. According to Matuscheck and Jansen [1] it means: Students encounter difficulties of ancient scientists, they use similar explanations and approaches of the ancient scientists, and are led by teachers to the ways of scientific thinking of today.

1.1 The Theory of Basic Matter

The ancient Greek philosophers have put a lot of thought into humanity and the world around them, they have come up with many recognized and accepted theories: Many of the current cultural and basic principles are based on these ancient Greek philosophies.

For instance, the basic questions that arose for these Greek philosophers are: "What is our world made of? What is the basic matter, material or substance? Just as important was the second basic thought that such basic matter must be eternal, that nothing can arise out of NOTHING and nothing can disappear to NOTHING – it is just the appearance of that basic matter which changes. With this realization, their attention was drawn to the following problems:

– the materials of the earth,
– the non-creationability and indestructibility of matter,
– the transformational ability of matter while retaining its basic substance" [3].

"Aristotle was the first to teach the difference between matter and property. This distinction of a thing being a 'carrier' of properties on the one hand, and on the other hand these properties themselves, were unknown to the Greek philosophers before Aristotle. Stemming from this knowledge, Aristotle discussed the theory that development and change, creation and destruction were nothing more than the transition from one essence to another" [4].

1.2 Transformation Concepts of the Alchemists

The age of the alchemists stretched from approximately the 4th to the 16th century, the Arabs being one of the main groups especially involved in this development. Alchemy, for them, was just another word for chemistry. This word is composed of the Arab word "al" and the Greek word "chyma" or metal production. This term shows the importance of metals for people and their wish to extract metals or even to transform non-noble metals into gold.

Many of the Arabs' writings even "provide directions for the artificial extraction of gold with the help of the 'Ferment of Ferments', the 'Elixir of Elixirs'. The correct mixture of the four elements is all-important and the 'spirit' (the heated, liquid quicksilver) has to permeate the 'matter' (lead, copper, etc.). The mysterious 'Elixir' itself is created through the correct fusion of the four elements, the body (metal) and the spirit (quicksilver), the male and the female. It assimilates these 'bodies' and colors them (therefore, they are known as 'tinctures') thereby transforming them to gold, causing them to multiply the amount thousand times" [5].

Also, the well-known scientist Albertus "believes in the possibility of creating gold, but he says, he knows of no alchemist who has succeeded in completely transforming metals" [5]. "Even up to the 18th century there was no lack of practical and witnessed alchemistic proof: there were examples of golden coins that supposedly were minted through alchemistic procedures, or nails made up of half of iron, and the other half of gold transformed from iron. Even court cases were sometimes ruled in favor of alchemistic operations, and last but not least, there were many swindlers who managed to show that through their 'successful transformations' it was indeed possible to transform metals into gold" [3].

In 1923, science was once again in a tizzy "when a Professor from a Berlin university reported that he had succeeded in transforming quicksilver into gold through treatment with electricity. The results were confirmed not only by various sides but there were several 'researchers' (even from Japan) who supposedly had made the same discoveries themselves. After a very thorough experimental examination, it was discovered a couple of years later that the minimal traces of gold came from the electrodes. This event finally banished the 'dream of gold' from the fantastic minds that had been pursuing it for centuries" [5].

1.3 The Phlogiston Theory

As a fuel like coal or petrol burns, one apparently can observe that they disappear. German scientist Stahl published his interpretation of this observation in 1697 and introduced the term Phlogiston (gr.: Phlox, the flame): "He started off with the combustion of sulfur and assumed that the sulfuric acid which was produced through the combustion act was sulfur bereft of its own burning principle" [6]. Stahl claimed, "every combustible and calcifiable material contains phlogiston. The combustion is a process through which the phlogiston is released by the substance. He assumed that air played a certain role assimilating the phlogiston and transferring it into leaves of trees. Through the process of reduction (by heating of calcified metal on charcoal), phlogiston was transferred from the coal into the heated substance. From then on, the processes of oxidation and reduction were recognized as being linked to each other. Proof was given through experiments of calcification of a metal through heat and the reduction through carbon" [3]:

$$\text{Metal} \rightarrow \text{Calcified Metal} + \text{Phlogiston}$$
$$\text{Calcified Metal} + \text{Phlogiston (in Carbon)} \rightarrow \text{Metal}$$

However, Stahl had to state that pure metal was a compound of metal and phlogiston, and calcified metal – known today as metal oxide – was expected an element. In addition to this came the fact that known measurements of masses did not agree with the theory: "He didn't place much value on the increase in weight of calcified metal; in the end, he tried to pawn this off as an assumed 'negative weight' of the phlogiston. The chemists were very one-sided focusing only on qualitative appearance, the Phlogiston Theory served this purpose well" [3].

"One can better understand the point of view of the Phlogiston Theory followers if one does not approach it through the material process but rather through an energy approach. One does not just see the flame escaping in the burning process; heat energy is also created. This aspect is completely ignored when one looks at the subject from a purely material approach. In order to be fair, and by looking at this from a physical-chemical process, one must see and replace the phlogiston as a kind of energy" [5]. Oxygen, which according to Empedocles was known as "fire-stuff" later on became "phlogiston" (Stahl) and "fire-air" (Scheele). "For several decades the term 'heat-stuff' was commonly used for all of the above names until the cause of the heat energy was discovered in the movement of small particles" [3].

Lavoisier clarified these facts through the synthesis and the decomposition of mercury oxide. He discovered the Law of Conservation of Mass and the Oxidization Theory. With these discoveries, the long era of alchemy and Phlogiston Theory was laid to rest and chemistry came forward as a leading scientific discipline.

1.4 Historic Acid–Base Theories

People throughout the ages were well familiar with the acidic taste of many fruits and vegetables. In addition, it has been known that alcoholic drinks produce vinegar if these are in an open container and have air contact. This vinegar had been used for thousands of years as a preservative. "The Latin word *acetum* for vinegar (actually, *acetum vinum*, means gone sour from wine) stands, from an etymological point of view, for *acer* = acrid and for *acidus* = acid or sour. Hence, the English word acid and the French word l'acide derive from this" [7].

Many attempts were made to explain what the acid taste was like. French scientist Lémery, from the 17th century, tried in a most unusual manner to explain the effect of acid through the idea of particles. He theorized that this invisible substance, which makes up the acidic taste, consists of moving spiky particles, which cause the acidic taste on the tongue [8].

In the beginning, there were only general explanations for the acidic taste phenomenon; for instance, indicators can show the acid manner by its color. Later, there were explanations of acidic–alkaline definitions based on the structure of the material – they were, however, always based on acidic or alkaline solutions as chemical substances. Today, the most common definition for beginners is the Broensted–Lowery concept and the idea of a proton transfer from one particle to another particle. However, this definition is based more on the function of ions or molecules as acid particles – and not on acids as substances! The different historic acid–base concepts are briefly described in the following periods of time:

BOYLE. In 1663, Robert Boyle characterized all acids by using the plant coloring, litmus: a red litmus color shows acidic solutions, a blue color basic

1.4 Historic Acid–Base Theories

solutions. Boyle became the creator of today's indicator paper. Apart from the color reaction, he also observed that acidic solutions are able to dissolve marble or zinc.

LAVOISIER. After the fall of the Phlogiston Theory and the discovery of oxygen, Lavoisier studied the combustion of carbon, sulfur and phosphorus in 1777. By dissolving the resulting non-metallic oxides in water, he found that all these solutions show acidic effects. Based on these examples, he defined acids as substances composed of a non-metal and oxygen. In addition, he found out that acids, combined with the "bases", the metal oxides, result in well-known salts: carbon dioxide forms the carbonates, sulfuric acid the sulfates, phosphoric acid the phosphates – the combination system of "acids and bases" was discovered!

DAVY. The discovery of the element chlorine by Davy in the year 1810, resulted in the finding of the gaseous compound, hydrogen chloride (HCl), and its watery solution, hydrochloric acid. With the realization that hydrogen chloride is essentially an oxygen-free compound, the search went on for a method of describing acid solutions in a new manner. A little later, the discovery of hydrogen sulfide (H_2S) and hydrogen cyanide (HCN) and their acidic solutions meant that eventually hydrogen was associated with the "principle of acids". However, hydrocarbons, a class of known hydrogen-containing compounds, do not possess acidic properties.

LIEBIG. Through the analysis of many organic acids and the knowledge of reactions of these solutions with non-noble metals to produce hydrogen, Liebig pragmatically stated in 1838: "Acids are substances that contain hydrogen which can be replaced by metals". The CH_3COOH molecule, for example, contain different H atoms, only the one H atom of the COOH group was defined as "hydrogen which can be replaced by metals". These new findings led to an immense progress in chemistry because diluted solutions of organic acids and those of common mineral acids held up to that definition.

ARRHENIUS. Upon examination of the electrical conductivity of many solutions, the term "electrolyte" for conducting substances was assigned. The acidic solutions also conducted electricity, and therefore belonged to the group of electrolytes. During additional freezing point depression experiments, it was discovered that electrolyte solutions showed a much greater effect than, for instance, sugar or ethanol solutions of the same concentration. Arrhenius was the first to interpret these experiences with "ions" and created the theory of electrolyte dissociation in aqueous solutions in 1884. In this way, the smallest particles of acidic solutions could be defined as hydrogen ions H^+(aq), and correlative remnant ions, the smallest particles of basic solutions as hydroxide ions OH^-(aq), and correlative remnant ions.

BROENSTED. After verifying the structure of atoms and ions by different models of nucleus and shell, hydrogen ions were classified as protons which do not exist freely and which connect with water molecules forming hydronium ions H_3O^+(aq). Based on this classification, Broensted and Lowery separately developed their own acid–base definition relating to protons in 1923. This definition proved independent of the aqueous solution and continued to expand more in the direction of particles (ions and molecules) rather than on substances: particles that give up protons (H^+ ions) to other particles are Broensted acids; they are also called proton donors. Particles that take protons are bases or proton acceptors.

Other definitions came next: concepts of Lewis, of Pearson and of Usanovich [7]. Because these concepts distanced themselves more and more from the classical acidic substances, they do not and should not play a significant role in the teaching–learning of chemistry for beginners.

1.5 "Horror Vacui" and the Particle Concept

Experiments with pipettes and wine siphons alerted the ancient natural philosophers to the fact that there are no air-free or other areas free of material on this earth, that as soon as a substance leaves a space it is replaced by another – mostly air. In this regard, Canonicus came up with a well-known formula, which states that "nature avoids empty space without any material, nature shows a horror vacui, a fear of empty spaces" [5].

Even Galileo Galilei knew of this phenomenon through the building of water wells and that it is impossible to pump water to the surface from a depth of over 10 m [5]. He attributed this depth as being the utmost power with which nature can prevent a vacuum. In 1643, Galilei created an experiment in order to measure the "resistenza del vacuo" or "resistance of vacuum" (see (a) in Fig. 1.1): The cylinder with a hook is filled with water, the moveable piston carries such a heavy mass that an empty space is created in the cylinder. One does not know "if this experiment was carried out or just described on paper" [5].

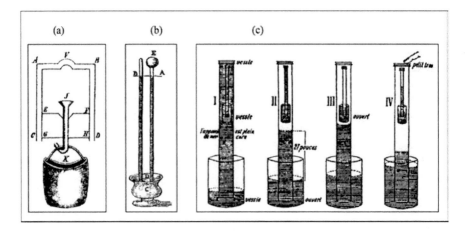

Fig. 1.1 Historical experiments to overcome the "horror vacui" [5]

This apparatus inspired his student Torricelli to replace the hard piston by mercury which is heavy and flows like a piston in a glass tube (see (b) in Fig. 1.1). With this experiment, first described in 1643, Torricelli was able to establish the rate of normal air pressure at 760 mm mercury. He was able to show the existence of a material-free zone: the vacuum [5] above the mercury column at the end of the glass tube.

Pascal established the final proof with his experiment "du vide dans le vide", the vacuum within the vacuum (see (c) in Fig. 1.1): in 1647 the Torricelli apparatus shows no evidence of gas pressure or no special "ether" above the mercury level [5]. Further experiments at various heights above sea level prove that the 20 km high column of air balances out a 760 mm elevation of a mercury column at sea level and by normal air pressure. This apparatus became useful for measuring air pressure, the first mercury barometer was built, and the unit of "1 mm Hg" got the unit 1 torr – due to the famous scientist Torricelli!

Because of this knowledge, Guericke developed efficient air pumps and demonstrated air pressure through spectacular experiments with the well-known "Magdeburger Halbkugeln": He took big half-spheres of metal which were joined together, he pumped them almost entirely free of air. Eight horses on one side and eight on the other pulled both half-spheres. It was only through using all their strength that the horses managed some times to overcome the air pressure and to separate the half-spheres with a big loud bang.

1.6 Atoms and the Structure of Matter

The old Greek philosophy offered at least two famous schools of thought. Some followers of Democritus and Leukipp were convinced that continual separation of a portion of matter must be finite and that matter contains atoms (gr.: atomos, indivisible). This idea postulates the concept of particles and the surrounding free space and was known as the **Hypothesis of Discontinuity**.

Aristotle and other philosophers claimed that the continual separation of matter was infinite. The idea of the impossibility of free space, which must separate particles from each other – the "horror vacui" – convinced them of the continual reconstruction of matter: **Hypothesis of Continuation.** The School of Aristotle had a huge influence and led to the suppression of the Hypothesis of Discontinuity for almost 2000 years [9].

After the vacuum was realized by the Torricelli experiments the "horror vacui" was hugely overthrown: one could conceptualize a vacuum. Gassendi developed on that base the particle concept from a sub-microscopic point of view, and in 1649 he rehearsed Democritus' idea of "atoms and empty space as the only principles of nature, apart from the complete full and complete empty space nothing else can be considered" [10]. After a two-thousand year interruption, scientists were finally able to embrace the Hypothesis of Discontinuity and to reconsider matter as being made up of smallest particles.

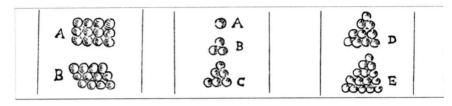

Fig. 1.2 Structural models of ice created by Kepler in the year 1611 [11]

Years before Gassendis' publications, Kepler through his observation of "six-sided snow crystals" reported in the year 1611: "There must be a reason for the fact that whenever the first snowflakes fall, they are always shaped in the form of a six-sided star, why don't they form a five-sided or even a seven-sided star?" [10]. Kepler assumed that steam contains "steam balls" and discussed condensation to snowflakes through various model-ball arrangements (see Fig. 1.2): "If one pushes similar-sized balls together so that they touch one another on a horizontal level they form either triangular or square shapes; a middle ball is surrounded by six other balls in the first model and four balls in the second model. The five-sided shape cannot provide a balanced coverage, and the six-sided shape can be reduced to a triangular form" [11].

In addition, Kepler spoke of the layering of closest packed balls in triangular patterns. He discovered the most compact ball set-up with coordination number of 12: "Again one ball is touched by 12 others – six neighboring balls at the same level, three on top, and three on the bottom". By assuming spheres as models of the smallest particles of water and in discussing their closest hexagonal order, Kepler was able to explain the permanent recurring six-sided snowflake shape. He thereby discovered the connection between the outer crystal shape and the inner arrangement of the smallest particles in the crystal, he postulated a first idea of the "chemical structure of ice".

Haüy [12] also contemplated arrangements of the smallest particles; however, he did not use balls as their models, but was thinking of "smallest portions of matter" and took "smallest crystals" as the same form of particles as in the big and visible crystal. Concerning the calcite crystal, he made the comment: "These rhomboids are what we would call integral particles (molécules intégrantes) of the calcite in order to differentiate them from the elementary particles (molécule élémentaires): one part is made of calcium, and the other part of carbon dioxide" [12].

Wollaston [13] went back to the ball-shape and backed it up with non-directed bonding of the smallest particles in crystal: "The existence of atoms merely requires mathematical points endowed with powers of attraction and repulsion equally on all sides, so that their extent is virtually spherical" [13]. In addition to this, he accented the closest packings of balls as models for crystal structures, but Wollaston was aware of the speculative nature: "It is probably too much to expect that we will ever find out the exact structure of any crystal" [13].

1.6 Atoms and the Structure of Matter

In 1808, from his observations, the English scientist Dalton was able to add a very important general theory on particles. In his work, "A New System of Chemical Philosophy" [14], he combined the idea of elements and the atomic concept and created the first table of atomic masses:

1. Every element consists of finite particles, the atoms.
2. Each atom of a given element has the exact same size and mass.
3. There are exactly as many types of atoms as there are elements.
4. Atoms can neither be created nor destroyed through chemical processes [14].

Dalton used different labeled spheres and, even if they were not always correct, came up with first models for compounds (see Fig. 1.3). He assumed, for instance, that the water molecule should be described by one H atom combined with one O atom, thinking of today's symbol HO (see Fig. 1.3, Number 21).

This incomplete concept delayed the correct Atomic Mass Table for quite some time. Later, with the concept of H_2O molecules, the suitable atomic masses were put in place. Many of Dalton's "elements like magnesia or lime" later proved to be compounds. Dalton's atomic theory was so fruitful that many organic substances could be successfully analyzed, especially with Liebig's combustion analysis. The only problem remaining was how to express the analytical results using chemical symbols, i.e. the different formulas for acetic acid molecules (see Fig. 1.4). For this reason, Kekulé initiated the Congress of Karlsruhe, which was held in 1861 in order to discuss the differentiation of atomic and molecular concepts. In other words, the first concepts regarding the structure of substances were defined as "atoms in elements" and "molecules in compounds". These definitions were just taken in textbooks on chemistry education until the middle of the 20th century.

Regarding the carbon–hydrogen compounds, Kekulé [16] published the theory of the connectivity of C atoms with four bonding units each and the structure of benzene molecules: "One can now assume that several carbon

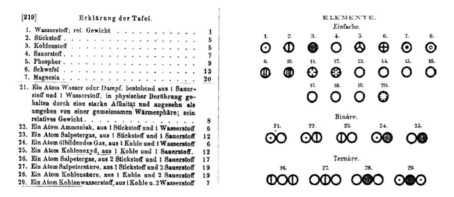

Fig. 1.3 Dalton's first Atomic Mass Table and first models in 1808 [14]

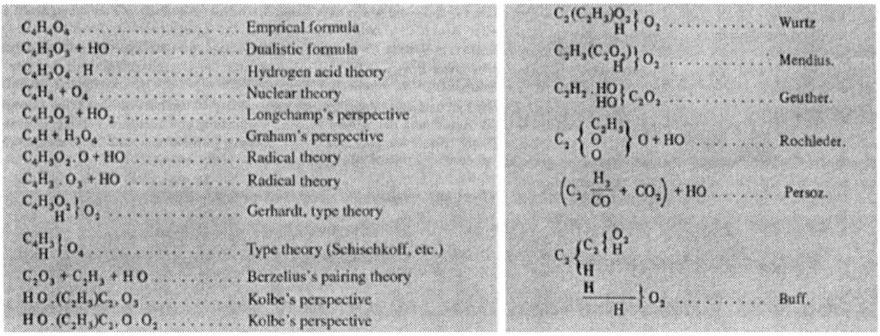

$C_4H_4O_4$	Empirical formula
$C_4H_3O_3 + HO$	Dualistic formula
$C_4H_3O_4 \cdot H$	Hydrogen acid theory
$C_4H_4 + O_4$	Nuclear theory
$C_4H_3O_2 + HO_2$	Longchamp's perspective
$C_4H + H_3O_4$	Graham's perspective
$C_4H_3O_2 \cdot O + HO$	Radical theory
$C_4H_3 \cdot O_3 + HO$	Radical theory
$C_4H_3O_3 \brace H \} O_2$	Gerhardt, type theory
$C_4H_4 \brace H \} O_4$	Type theory (Schischkoff, etc.)
$C_2O_3 + C_2H_3 + H \cdot O$	Berzelius's pairing theory
$HO \cdot (C_2H_3)C_2 \cdot O_3$	Kolbe's perspective
$HO \cdot (C_2H_3)C_2 \cdot O \cdot O_2$	Kolbe's perspective
$C_2(C_2H_3)O_2 \brace H \} O_2$	Wurtz
$C_2H_3(C_2O_2) \brace H \} O_2$	Mendius
$C_2H_7 {\cdot HO \atop HO} \} C_2O_2$	Geuther
$C_2 \left\{ {C_2H_3 \atop O \atop O} \right\} O + HO$	Rochleder
$\left(C_2 {H_3 \over CO} + CO_2 \right) + HO$	Persoz
$C_2 \left\{ {C_2 \brace H} {O_2 \atop H} \right.$	
${H \over H} \} O_2$	Buff

Fig. 1.4 Mental models of acetic acid molecules in the year 1861 [15]

atoms form a chain and are connected through bonding units. Furthermore, the connectivity always alternates through one or two bonding units. If we assume that the two C atoms, which close the chain, are connected through a single bonding unit, we will obtain a closed ring, which offers another six bonding units" [16]. Chemical symbols like C_6H_6 were established bit by bit. Only through overcoming the Oscillation Theory for single and double bonds in benzene molecules, the ring symbol for the benzene molecule could be developed: this is the most common symbol today.

After initial ideas of atoms and molecules, there was only one remaining kind of particle missing in order to complete the basic kit: ions. Arrhenius [17] observed variations on the electrical conductivity of different salt solutions and related his observations to decreasing freezing points and increasing osmotic pressures compared to usual solutions of sugar or ethanol. In 1884, he postulated the Dissociation Theory and the ions as being the smallest particles of all solutions of salts, acids and bases. Initially, this first concept for ions was not really understood by his colleagues, so that Arrhenius stated: "How can one think of free sodium in a sodium chloride solution knowing how sodium behaves in water, and that free chlorine should exist in the colorless and odorless sodium chloride solution knowing that a chlorine water solution is yellow and smells strongly... The fact that I believed in the destruction of salt molecules to form ions was considered a problem and opposed by members of the faculty. Knowing that most chemists would strongly reject, I tried to avoid emphasizing the Dissociation Theory. This delayed the publication of the theory for three additional years" [18]. It took a long time to introduce the idea of ions in chemistry lectures and textbooks, and misconceptions regarding "salt molecules" continued to exist way into the 20th and 21st century.

In 1912 the German scientist Laue [19] discovered the new interference pattern through X-ray radiation of crystals and finally confirmed that molecules are not the smallest particles of salt crystals: "The term molecule no longer plays a role regarding NaCl and KCl. Each Cl atom touches six metal atoms with an exact distance between them" [19]. However, it became obvious that the

1.6 Atoms and the Structure of Matter

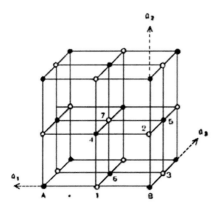

Fig. 1.5 Structural models of sodium chloride by Laue [19] and Bragg [20]

salt molecule model no longer made sense especially since Laue's model of the sodium chloride structure became more and more popular (see Fig. 1.5).

The English researcher Bragg [20] stated this also: "For instance, chemists had talked of common salt, sodium chloride, as being composed of 'molecules' of NaCl. My very first determination showed that there are no molecules of NaCl consisting of one atom of sodium and one of chlorine. The atoms are arranged like the black and white squares of a chessboard, though in three dimensions. Each atom of sodium has six atoms of chlorine around it at the same distance, and each atom of chlorine has correspondingly six atoms of sodium around it" [20]. He even constructed a spherical packing model for his concept (see Fig. 1.5).

Roelleke [21] discussed the historical development of structural X-ray analysis and, using an X-ray machine for classrooms, he obtained interference patterns according to Laue and Debye–Scherrer. He proposed ways to introduce X-ray analysis in chemical education.

Until way into the middle of the 20th century, one was not willing to let go of the trusted concepts of salt molecules. Bragg was even asked by other colleagues to save the molecule concept: "Some chemists at that time were very upset indeed about this discovery and begged me to see that there was just a slight resemblance of one atom of sodium to one of chlorine as a properly married pair" [20].

Even up until now, the misconceptions of molecules in salt crystals are so strongly anchored that even in scientifically trained circles one is not using the correct mental model of ions or ion grids according to salt crystals. The historical development of models concerning structure of matter and the big number of misconceptions can be read more precisely in other places ([22, 23, 24]) and in Chap. 5.

Conclusion: When we consider "historical misconceptions" and "scientific errors", which intelligent scientists have produced over the ages in their attempts to discover how matter is structured, we must show appreciation to our students who may, in their attempts, stumble across similar misconceptions as in the history. For this reason, it is important that each educator should be aware of the history in chemistry and physics in order to appreciate mental models of students today. It may also be appropriate to teach first the historical ways of discovering particles and chemical structures, before the most valid concepts of today are introduced.

References

1. Matuschek, C., Jansen, W.: Chemieunterricht und Geschichte der Chemie. Praxis (Chemie) 34 (1985), 3
2. Wandersee, J.H.: Can the history of science help science educator anticipate student's misconceptions? Journal of Research in Science Teaching 23 (1985), 581, 594
3. Strube, W.: Der historische Weg der Chemie. Leipzig 1976
4. Reuber, R., u.a.: Chemikon – Chemie in Übersichten. Frankfurt 1972
5. Lockemann, G.: Geschichte der Chemie. Berlin 1950
6. Bugge, G.: Das Buch der Grossen Chemiker. Band 1. Weinheim 1955
7. Hammer, H.O.: Saeure-Base-Vorstellungen. Geschichtliche Entwicklungen eines Begriffspaares. PdN-Ch. 44 (1995), 36
8. Haeusler, K.: Highlights in der Chemie. Köln 1998 (Aulis)
9. Dijksterhuis, F.J.: Die Mechanisierung des Weltbildes. Berlin 1956
10. Lasswitz, K.: Geschichte der Atomistik. Baende 1 und 2. Hamburg 1890
11. Kepler, J.: Vom sechseckigen Schnee. Frankfurt 1611
12. Hauy, R.J.: Traité élémentaire de physique (1784). Übersetzung von Blumhoff: Weimar 1804
13. Wollaston, W.H.: On the elementary Particles of Certain Crystals. Philosophical Transaction of the Royal Society of London 103 (1813), 51
14. Dalton, J.: Über die Absorption der Gasarten durch Wasser und andere Fluessigkeiten (1803), A new System of Chemical Philosophy (1808). In: Ostwalds Klassiker Nr. 3, Leipzig 1889
15. Walther, W.: Chemische Symbole in der Vergangenheit und Gegenwart. CU 13 (1982), H.2
16. Kekulé, A.: Ueber die Constitution und die Metamorphosen der chemischen Verbindungen und ueber die chemische Natur des Kohlenstoffs (1958), Untersuchungen über aromatische Verbindungen (1866). In: Ostwalds Klassiker Nr. 145, Leipzig 1904 (Engelmann)
17. Arrhenius, S.: Galvanische Leitfaehigkeit der Elektrolyte (1883). In: Ostwalds Klassiker Nr. 160, Leipzig 1907 (Engelmann)
18. Bugge, G.: Das Buch der Grossen Chemiker, Baende 1 und 2, Weinheim 1955
19. Laue, M.: Die Interferenz der Roentgenstrahlen. In: Ostwalds Klassiker Nr. 204, Leipzig 1923
20. Bragg, L.: The history of X-ray Analysis. London 1943 (Longmans)
21. Roelleke, R., Barke, H.-D.: Max von Laue: ein einziger Gedanke – zwei große Theorien. PdN-Chemie (1999), Heft 4, 19
22. Barke, H.-D.: Chemiedidaktik zwischen Philosophie und Geschichte der Chemie. Frankfurt 1988 (Lang)
23. Barke, H.-D.: Max von Laue – ein Experiment verifiziert zwei große Theorien. In: Barke, H.-D., Harsch, G.: Chemiedidaktik Heute. Heidelberg 2001 (Springer)
24. Hilbing, C.: Alternative Schuelervorstellungen zum Aufbau der Salze als Ergebnis von Chemieunterricht. Muenster 2003 (Schueling)

Chapter 2
Students' Misconceptions and How to Overcome Them

Misconceptions are not only to be observed in today's children or students – even scientists and philosophers developed and lived with many misconceptions in the past (see Chap. 1). Historical concepts and their changes are very interesting because similar ideas can help our students today: just like early scientists did they develop their own concepts by similar observations e.g., in regard to combustion. Ideas that are developed without having any prior knowledge of the subject are not necessarily wrong but can be described as **alternative**, **original** or **preconcepts** [1]. Every science teacher should know these preconcepts for his or her lessons – this is why many empirical researchers are working all over the world.

Increasingly however, researchers are also finding chemical misconceptions in advanced courses. Because they cannot be only attributed to the students but mainly caused by inappropriate teaching methods and materials, they can be called **school-made misconceptions**. They are clearly different from preconcepts that tend to be unavoidable. Inappropriate teaching methods can be stopped by keeping teachers up-to-date in their subject through advanced education.

One should attempt to find as many preconcepts and school-made misconceptions and discuss them with pre-service and in-service teachers. Another important task is to make suggestions of instructional **strategies to improve lessons,** which will lead to challenge preconceptions and school-made misconceptions: recommending alternative strategies to the traditional approaches, setting up convincing laboratory experiments, using more structural models or new technology-based methods etc.

2.1 Students' Preconcepts

Self-developed concepts made by students do not often match up with today's scientific concepts. One fails to take into account that these young folks have often, through observation, come up with their own mostly intelligent ideas of the world. In this sense, they are in good company considering that ancient scientists and natural philosophers also used their power of observation and logic in order to shape their ideas. Often, these scientists and philosophers did not use additional experiments to back up their theories (see Chap. 1).

When students talk about combustion, saying that "something" disappears and observe that the remaining ash is lighter than the original portion of fuel, then, they have done their observation well and have come up with logical conclusions. This is why we cannot describe their conclusions as incorrect but rather as:

- original or pre-scientific ideas,
- students preconceptions or alternative ideas,
- preconcepts.

It is common to come across several preconcepts at the beginning stages of scientific learning at the elementary, middle and high school levels of chemistry, biology and physics. Before conclusions are systematically made regarding the important issues of chemistry in the following chapters, three general examples of a student's **preconcepts** will be presented:

- the sun revolves around the earth,
- a puddle is sucked up by the sun's rays,
- the wood of a tree comes from the soil.

Sun and Earth. Most children's first experiences regarding the sun are accompanied by comments made by their families and neighbors: "Look, the sun will rise in the morning, at midday it will be at its highest point and in the evening it will set". Observations regarding sunrise, sunset, its own cycle and the common manner of speech regarding this subject must lead the child to the idea: "The sun cycles around the earth". In some of her interviews, Sommer [2] even comes across the idea of the earth as being a disc: "Children imagine the earth to be a disc over which the sky stretches parallel. The sun, the moon and the stars are to be found in the sky; there is no universe" [2].

Greek natural philosophers developed their ideas 2000 years ago. Ptolemy especially imagined the earth to be at the center of everything and pondered: "The sun moves around the earth". It was at the end of the 16th century that Copernicus, after exact observation of the movement of the planets, came up with the heliocentric image of the earth: "The earth is one of the sun's many planets, like these planets, the earth is revolving in a particular pathway around the sun and it also revolves on its own axis". Considering the uproar of the church at that time and the ensuing Inquisitions, one can imagine how stable Ptolemy's theory was present in the minds of people of the time. It was the real wish of the church to keep people in this ignorance: The earth was supposed to be the center of the universe.

Children and adolescents often, through their own observations, come up with similar concepts like Ptolemy, of course – there is no way to make discoveries like Copernicus' and to develop the heliocentric view of the earth. Teachers have to use the best methods and technology, e.g. a planetarium, in order to convince the kids to free themselves from their original ideas and to accept that the earth is revolving around the sun.

In order to have convincing lessons, it is important that young people have enough opportunities to first express and compare their ideas of the universe. Only after children feel uncomfortable with their ideas, the new and current worldview should be introduced. The children should realize that their view of the world is also quite common and even scientists in the past believed that "the sun moves around the earth". Good teaching with models like moving spheres in a planetarium should finally convince children of the revolving earth.

Puddles and Sun Rays. Through conversations with elementary school children regarding the disappearance of puddles on a sunny day, it is obvious that they believe that the sunrays "soak up the water", that "water disappears to nothing". When asked, many teachers admit that they find this explanation "cute" and often do not bother to correct or discuss it: they let the children be with their "sunray theory" and their view of the "elimination of water".

If, on the other hand, the teachers would carry out experiments showing the vaporization of water and the resulting condensation of the steam to liquid water, the scientific view could be started. If one also introduces the idea of particles and the mental model of increasing movement of the water particles through heat, a child would much better understand that the water particles mixes with air particles and therefore remain in the air.

They, furthermore, would understand that particle movement and diffusion of energy-rich particles are responsible for the evaporation of water. This would also lead the children to a logical understanding of the conservation of mass for later science lessons and understanding chemical reactions, especially regarding combustion. It is necessary however, that children can express their own view about the "disappearance of water" before they learn the scientific concept. To be convinced by the scientific concept they should look to demonstrated or self-done experiments and compare with their own view. Following these discussions, after more experiences with evaporation and condensation of water, children or students may realize their conceptual change (see Sect. 2.3).

Wood and Earth. "When people are given a piece of wood and asked how the material got into the tree they commonly reply that most of it came from the soil" [3]. Even though, in biology, the subject of photosynthesis is taught with the use of carbon dioxide, water, light and heat for the synthesis of sugar and starch, still many students when asked where wood comes from, reply: "from the soil". Most students seem to have their knowledge of biology lectures in special "compartments" of their brain. They do not link them to their every-day life understanding: "Presumably most of the graduates would have been able to explain the basics of photosynthesis (had that been the question), but perhaps they had stored their learning about the scientific process (where carbon in the tree originates from gaseous carbon dioxide in the air) in a different compartment from their 'everyday knowledge' that plants get their nutrition from the soil" [3].

This example should indicate that preconcepts can even still be used for a subject when the related lectures have dealt with the appropriate scientific idea. When one forgets or deliberately avoids making connections between this newly attained knowledge and well-established observations, the new scientific knowledge will not stay stable – the learner is going back to his or her previous preconcepts: both, preconcept and scientific thinking are stored in "compartments", in separate areas of the cognitive structure.

Teachers cannot automatically assume that in a particular lesson any preconceptions regarding this lesson will appear. It is necessary to diagnose such concepts and, in the case of misconceptions, to plan a lesson which integrates new information with these concepts. If the lesson is about photosynthesis it would be advisable to bring in everyday aspects, that wood is made up of carbon dioxide and water steam from the air, made up of carbon dioxide and water molecules. One could emphasize that plants need the earth in order to transport minerals from the roots to the branches but that, as hard as it is to believe, the solid and massive wood develop due to chemical reactions of colorless gases. Again, one could point out that even ancient scientists believed in the historical humus theory and could not understand when the German Justus von Liebig experimentally verified photosynthesis in the middle of the 19th century.

2.2 School-Made Misconceptions

When students get involved in a subject matter that is more difficult, a different type of problem arises: school-made misconceptions. Due to their complexity, it is not often possible to address certain themes in a cut-and-dry manner. Despite competent and qualified teachers, occasionally questions remain open and problems are not really solved for a full understanding: school-made misconceptions develop. A few examples should illustrate this.

Composition of Salts. A famous example of school-made misconceptions of our students arises from the Dissociation Theory of Arrhenius. In 1884, he postulated that "salt molecules are found in solid salts as the smallest particles and decompose into ions by dissolving in water". Later, with the concept of electrons, the misconception that "atoms of salt molecules form ions through electron exchange" was born. Today, experts recognize that there are no salt molecules, that ions exist all the time – even in the solid salt. By dissolving the solid salt, water molecules surround the ions, and hydrated ions are not connected, they move freely in the salt solution.

Amazingly one can observe that even today – in the year 2004 – the historic misconceptions are quite common: "Sodium chloride consists of sodium and chlorine atoms. Each chlorine atom takes an electron from the sodium atom so the chlorine atom will have a negative electrical charge, the sodium atom a positive one" [4]. A magazine for young students – published in the year 2004 [5] – contains the same misconceptions (see Fig. 2.1).

2.2 School-Made Misconceptions

Fig. 2.1 Today's misconceptions about common salt and salt solution [5]

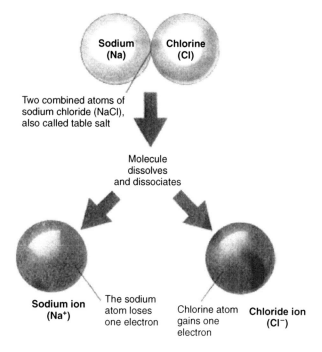

Also in the related subject of chemical bonding, one elaborates mostly on electron-pair bonding and only briefly on ionic bonding. The result is that students will not have any lasting concept of ions in an ion lattice. Regarding the question which particles are found in mineral water which contains calcium chloride, many students answer "Cl-Ca-Cl molecules" [6]. In this case, misconceptions have been developed during lessons – these misconceptions are school-made! Such misconceptions even occur if ions in the recommended issue of electrolysis of salts have not been correctly taught [7]. In the cited publications and in the following chapters, suggestions and ways in teaching the issue of ions and ionic bonding in a more successful and effective manner will be presented.

Chemical Reactions. It is traditional in chemistry lessons to separate chemical reactions from physical processes. The formation of metal sulfides from its elements by releasing energy is described in every case as a chemical reaction. In contrast, the dissolving of substances in water is often regarded as a "physical process" because matter "does not actually change", the dissolved substance can be regained in its original form through "physical" separation procedures. If one takes sodium hydroxide and dissolves it in a little water, a colorless solution appears and releases heat; the solution conducts electricity and produces a high pH value. Critical students regard this solution as being a new material and the production of heat shows an exothermic reaction. From this example one can see that it does not make any sense to separate the transformation of matter into "chemical" and "physical" processes [8]. If we routinely continue to do this in the sense of "we've always done it this way", automatic school-made misconceptions would arise based on teaching traditions in school.

Composition of Water. "Water is composed of hydrogen and oxygen" [1] – one often hears these or similar statements in classrooms about compounds, which supposedly "contain" certain elements. These expressions arise from a time when it was common to analyze and find out which elements make up certain compounds. Insiders know the background of these statements – for novices however, they will lead to school-made misconceptions: students would associate the substances copper and sulfur in the black copper sulfide, particularly as experiments show that one can remove these elements out of copper sulfide. It would be better, in introductory classes, to point out that the metal sulfides could be produced from metals and sulfur or to show that one can obtain the elements from the compound. Later on, if one is aware of "atoms" and "ions" as the smallest particles of matter, one can expand on these statements, that the compound "contains" special atoms or ions, that one water molecule contains two H atoms and one O atom connected and arranged in a particular spatial structure. But the pure sentence "water contains hydrogen and oxygen" will develop school-made misconceptions!

2.3 Students' Concepts and Scientific Language

One should be aware that newly acquired concepts are not sustainable forever and can be easily affected when lessons are over. Concepts regarding life in general, which have been sustained over several years, are more deeply rooted than new concepts, which have more recently been picked up in lessons. It is therefore necessary to repeat and intensify these newly "acquired" concepts in order to anchae them in the minds of students.

Teachers should also be aware that students will be insecure when discussing these new scientific concepts with friends or relatives – they will resort to slang or every-day language. Although they know about conservation of mass they will have to deal with terms like "the fuel is gone" or "spots are removed" [1]. One should try to help students begin to reflect on the use of such every-day language. Then, they could discuss these thoughts with friends or relatives – in this sense, they would become competent and improve the much wished ability to be critical. Such abilities could certainly have a positive effect on society in that such scientific knowledge would not only be discussed in colloquial terms, but that the students could competently use the proper terminology and pass it on to friends and family.

Many school-made misconceptions occur because there are problems with the specific terminology and the scientific language, specially involved substances, particles and chemical symbols are not clearly differentiated. If the neutralization is purely described through the usual equation, $HCl + NaOH \rightarrow NaCl + H_2O$, then the students have no chance to develop an acceptable mental model that uses ions as smallest particles.

2.3 Students' Concepts and Scientific Language

When questioned, on which the neutralization reaction is based, students mostly come up with mental models of H-Cl molecules and of Na-O-H molecules. If one would discuss both ion types in hydrochloric acid and sodium hydroxide solutions, and if one would sketch them in the form of model drawings [9], it would probably be possible for the students to develop the right mental model and scientific language at this level (see Chap. 7). This would also enable them to interpret the equation for the above reaction with the help of ionic symbols.

Johnstone [10] elucidated this connection (see Fig. 2.2): "We have three levels of thought: the macro and tangible, the sub-micro atomic and molecular, and the representational use of symbols and mathematics. It is psychological folly to introduce learners to ideas at all three levels simultaneously. Herein lay the origins of many misconceptions. The trained chemist can keep these three in balance, but not the learner" [10]. Specially Gabel [11] points out that teachers like to go from the macro level directly to the representational level and that students have no chance of following this concept: "The primary barrier to understanding chemistry is not the existence of the three levels of representing matter. It is that chemistry introduction occurs predominantly on the most abstract level, the symbolic level" [11].

It appears to be particularly difficult even at secondary schools to make this transition from the macroscopic level directly to the representational level. This, again, leads to school-made misconceptions, students are mixing substances from the macroscopic level with particles from the sub-micro level: "hydrochloric acid is giving one proton" (instead of "one H_3O^+(aq) ion gives one proton"), "oxygen takes two electrons" (instead of "one O atom is taking two electrons"). On the one hand, the students do not see any connection between both levels, on the other hand it is left up to them to figure out which mental model they may choose concerning the sub-microscopic level: they are building up ideas on their own, mostly wrong ones.

The misconception concerning the neutralization example above could be avoided if, after carrying out the experiment, one would describe the observations at the **macro level**. By interpreting these observations, one could ask

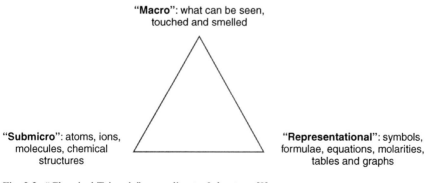

Fig. 2.2 "Chemical Triangle" according to Johnstone [9]

questions regarding the particles related to the reaction. These could be answered using ions and ionic symbols at the **sub-micro level**. It would be even better if one used model drawings related to the hydrated ions in hydrochloric acid and in sodium hydroxide solution [9]. Only when the reaction of H^+(aq) ions with OH^-(aq) ions to form H_2O molecules has been made clear on the sub-micro level, the **representational level** and the chemical symbols will be successfully attained. On this level other reaction equations may be written or related calculations could be done.

2.4 Effective Strategies for Teaching and Learning

"All teaching should begin with children's experiences – each new experience made by children in a classroom is organized with the aid of existing concepts" [12]. "Without explicitly abolishing misconceptions it is not possible to come up with scientific sustainable concepts" [13]. "Lessons should not merely proceed from ignorance to knowledge but should rather have one set of knowledge replace another. Chemical education should be a bridge between students' preconcepts and today's scientific concepts" [14].

These statements make it quite obvious that teachers should not assume their students enter their classroom with no knowledge or ideas whatsoever. A lesson, which does not take into account that students have existing concepts, usually enables them to barely following the lecture until the next quiz or exam. After that, newly acquired information will gradually be forgotten: students tend to return to their old and trusted concepts.

Nowadays, teachers and pedagogy experts agree that one should be aware of student's ideas before the "bridge can be successfully made between the preconcepts and the scientific ones" [14]. Therefore, an important goal is to allow students to express their own preconcepts during a lesson or, in the attempt to introduce new subject matter in a lesson, to let them be aware of inconsistencies regarding their ideas and the up-to-date scientific explanation. In this way, they can be motivated to overcome these discrepancies. Only when students feel uncomfortable with their ideas, and realize that they are not making any progress with their own knowledge will they accept the teacher's information and thereby build up new cognitive structures.

For the teaching process, it is therefore important to take students' developmental stages into account according to:

- student's existing discrepancies within their own explanations,
- inconsistencies between preconcepts and scientific concepts,
- discrepancies between preliminary and correct explanations of experimental phenomena,
- possibilities of removing misconceptions,
- possibilities of constructing acceptable and skilled explanations [15].

2.4 Effective Strategies for Teaching and Learning

One should especially take into consideration that, regarding constructivist theories, it is only possible to change from preconcepts to scientific concepts if

- individuals are given the chance to construct their own learning structures,
- each student can get the chance to actively learn by himself or herself,
- "conceptual growth" can occur congruent to Piaget's assimilation, or even
- "conceptual change" can occur congruent to Piaget's accommodation [15].

If a student does not believe that "sunrays absorb a puddle" (see Sect. 2.1), he or she can then, using the particle model of matter with the idea of moving particles, successfully develop a scientific concept about the evaporation of water. There is an extension of the already established particle concept taught in lessons before – a **conceptual growth** appears.

Should yet another student believe that "sunrays soak up the puddle" (see Sect. 2.1), perhaps through having learned it at the elementary school, then he or she is unlikely to want to let go of this concept. Even if lessons about the particle model of matter are plausible and logical, he or she is unlikely to integrate it or to swap it against the "sun's absorption ability". If the teacher helps to understand the scientific concept through the introduction of self-moving particles, then this student has to take a huge step in releasing his old ideas: a **conceptual change** has to develop in his cognitive structure. To push this development to a new mental model it would be advantageous to do his or her own active experiments and model drawings according to the particle model of matter and self-moving particles (see Chap. 4).

Also the advancement from a destructive concept to a preservation concept – e.g., concerning the combustion or metal-oxygen reactions – would lead to such a change in the cognitive structure, to a conceptual change.

Taber came up with the picture of a "**Learning Doctor**" as a means of discovering individual misconceptions and a suitably-related science class regarding conceptual growth or conceptual change [3]: "A useful metaphor here might be to see part of the role of a teacher as being a learning doctor: (a) diagnose the particular cause of the failure-to-learn; and (b) use this information to prescribe appropriate action designed to bring about the desired learning. Two aspects of the teacher-as-learning-doctor comparison may be useful. First, just like a medical doctor, the learning doctor should use diagnostic tests as tools to guide action. Secondly, just like medical doctors, teachers are 'professionals' in the genuine sense of the term. Like medical doctors, learning doctors are in practice (the 'clinic' is the classroom or teaching laboratory). Just as medical doctors find that many patients are not textbook cases, and do not respond to treatment in the way the books suggest, so many learners have idiosyncrasies that require individual treatment" [3].

In a project in progress **Barke and Oetken** agree to diagnose preconcepts and school-made misconceptions, but in addition they will integrate them into lectures to develop sustainable understanding of chemistry [16]. For the past 20–30 years educators continue to observe nearly the same misconceptions of students, therefore they assume that related lectures at school are not changing much. Hence, being convinced that preconcepts and school-made misconceptions have to be discussed in chemistry lectures, there are two hypotheses to influence instruction:

1. One should first discuss the misconceptions and come up with the scientific explanation afterwards, 2. one instructs the scientific concept first and afterwards students compare it with their own or other misconceptions from literature.

Oetken and Petermann [17] use the first hypothesis for their empirical research concerning the famous preconcept of combustion: "Something is going into the air, (...) some things disappear". In their lectures they showed the burning of charcoal and discussed alternative concepts like: "charcoal disappear some ashes remain". Afterwards they used the idea of a cognitive conflict: little pieces of charcoal are deposited in a big round flask, the air is substituted by oxygen, the flask is tightly closed and the whole thing is weighed using analytical balance. Pressing the stopper on the flask and heating the area of the charcoal, the pieces ignite and burn until no charcoal remains. The whole contents are weighed again, the scales afterwards present the same mass as before.

Working with this cognitive conflict the students find out that there must be a reaction of carbon with oxygen to form another invisible gas. After testing this gas by the well-known lime water test one can derive: the gas is carbon dioxide. Presenting misconceptions first and instructing the scientific concept afterwards can enable students to compare and investigate by themselves what is wrong with statements like "some things disappear" or "combustion destroys matter, mass is going to be less than before". Integrating preconcepts in lectures by this way will improve sustainable understanding of chemistry; by comparing misconceptions with the scientific concept students will internalize the concept of combustion. More results in line with this hypothesis will come up in the future.

Barke, Doerfler and Knoop [18] planned lectures according to the second hypothesis in middle school classes: 14–16 years old students were supposed to understand acids, bases and neutralization. Instead of taking the usual equation "HCl + NaOH $\rightarrow$ NaCl + H_2O" for the reaction, H^+(aq) ions for acidic solutions and OH^-(aq) ions for basic solutions were introduced, the ionic equation of the *formation of water molecules* was explained: "H^+(aq) ions + OH^-(aq) ions $\rightarrow$ H_2O molecules". Later it was related that, with regard to neutralization, other students think of a "*formation of salt*" because "NaCl is a product of this neutralization". Students discussed this idea with the result that no solid salt is formed by neutralization, Na^+(aq) ions and Cl^-(aq) ions do not react but only remain by the neutralization. These ions are therefore often called "spectator ions".

So students were first introduced to the scientific idea of the new topic, and afterwards confronted with well-known misconceptions. By comparing the scientific idea and the presented misconceptions the students could intensify

2.4 Effective Strategies for Teaching and Learning

the recently gained scientific concept. Preliminary data show that this hypothesis is successful in preventing misconceptions concerning the neutralization reaction. There will be more empirical research as to whether this method is the most sustainable strategy for teaching and learning.

With regard to teaching about ions and ionic bonding **Barke, Strehle and Roelleke** [19] evaluated lectures in the sense of hypothesis two: by the introduction of "atoms and ions as basic particles of matter" based on of Dalton's atomic model (see Fig. 5.10 in Chap. 5) scientific ideas according to chemical structures of metal and salt crystals are reflected upon.

Using this method of instruction, all questions regarding chemical bonding are reduced to undirected electrical forces surrounding every atom or ion – no electrons or electron clouds are involved at this time. However, the **structure of elements and compounds** can be discussed because spatial models or model drawings are possible based on Dalton's atomic model (see Fig. 2.3). In the first two years of chemical education only the structure of matter should be considered (see Chaps. 4 and 5) – the detailed questions regarding chemical bonding should be answered later after the instruction of the nucleus-shell model of the atom or the ion. By combining ions to ionic lattices in salt structures students learn the scientific idea about the composition of salts: cations and anions, their electrical attraction or repulsion, their arrangement in ionic lattices. Through this strategy of combining ions and using ion symbols it is possible to prevent most of the related global misconceptions (see Fig. 2.1) (Chap. 5)!

Last but not least **Barke and Sileshi** [20] formulated a "Tetrahedral-ZPD" chemistry education metaphor as another framework to prevent students' misconceptions (see Fig. 2.4). "If chemical education is to be a discipline, it has to have a shape, structure, clear and shared theories on which testable hypotheses can be raised. At present we are still in some respects dabbling in chemistry education alchemy, trying to turn lead into gold with no clear idea about how this is to be achieved. Some factions proclaim a touchstone in some pet method such as Problem-Based Learning or Computer-Assisted Learning or Multimedia Learning or Demonstration, while others are dabbling in

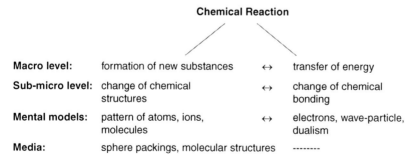

Fig. 2.3 Chemical structure and bonding with regard to the chemical reaction [1]

Fig. 2.4 Tetrahedral-ZPD chemistry education metaphor [20]

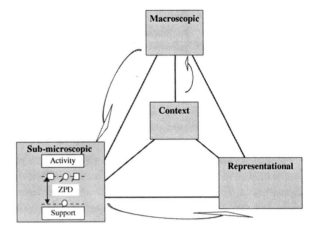

Conceptual Assessment, Microscale Labs, and fancy textbooks accompanied by teachers' guides. None of these things are bad, but what theory is driving them? Is there any evidence that they are achieving what they aim for? Are we any nearer to making gold?" [21].

To respond to such calls Sileshi and Barke – after reviewing the major chemistry education concepts – proposed the Tetrahedral-ZPD metaphor. This metaphor re-hybridizes the very powerful 3D-tetrahedral chemistry education concept proposed by Mahaffy [22]: macroscopic, molecular, representational, and human element. With the idea of the "Zone of Proximal Development (ZPD)" of social constructivist Vygotsky [23], ZPD should describe "the distance between the actual development level as determined by independent problem solving and the level of potential development as determined through problem solving under adult guidance or in collaboration with more capable peers" [23].

The basic elements of this metaphor are what Shulman [24] has labeled "Pedagogical Content Knowledge (PCK)" integrated with contextual and research knowledge: "Pedagogy-Content-Context-Research Knowledge (PCCRK). Content knowledge refers to one's understanding of the subject matter, at macro-micro-representational levels; and pedagogical knowledge refers to one's understanding of teaching-learning processes; contextual knowledge refers to establishing the subject matter within significant social-technological-political issues; and research knowledge refers to knowledge of 'what is learned by student?', that is, findings and recommendations of the alternative conceptions research of particular topics in chemistry" [24].

Sileshi and Barke further conduct an empirical research to evaluate the effects of the Tetrahedral-ZPD Metaphor on students' conceptual change (see Fig. 2.4). Knowing that high school students in Ethiopia mostly memorize chemical equations without sufficient understanding, that they are not used to thinking in models, or developing mental models according to the structure of matter, new teaching material and worksheets for the application of the particle model of matter and Dalton's atomic model are prepared.

In pilot studies lasting for six weeks the research was carried out with an experimental-control group design: pre-tests and post-tests were used to collect data before and after the intervention. First results from the post-tests indicated that the students in the experimental group, taught with the new teaching material according to the structure of matter, show significantly higher achievement compared with the students in the control group: students' misconceptions in the experimental group after they were taught using the new teaching material based on Tetrahedral-ZPD, are less than in the control group. The main studies will follow.

References

1. Barke, H.-D.: Schuelervorstellungen. In: Barke, H.-D., Harsch, G.: Chemiedidaktik Heute. Lernprozesse in Theorie und Praxis. Heidelberg 2001 (Springer)
2. Sommer, C.: Wie Grundschueler sich die Erde im Weltraum vorstellen – eine Untersuchung von Schuelervorstellungen. ZfDN 8 (2002), 69
3. Taber, K.: Chemical Misconceptions – Prevention, Diagnosis and Cure. Volume I. London 2002 (Royal Society of Chemistry)
4. Gerlach, H.: Salz auf unserem Steak. Sueddeutsche Zeitung, Magazin. Stuttgart 2004, 34
5. Welt der Wissenschaft (deutsche Übersetzung). Bath, England, 2004 (Parragon), 34
6. Barke, H.-D., Selenski, T., Sopandi, W.: Mineralwasser und Modellvorstellungen. PdN-ChiS 52 (2003), H.2, 15
7. Hilbing, C., Barke, H.-D.: Ionen und Ionenbindung: Fehlvorstellungen hausgemacht! Ergebnisse empirischer Erhebungen und unterrichtliche Konsequenzen. CHEMKON 11 (2004), 115
8. Barke, H.-D., Schmidt, F.K.: Chemische Reaktionen und physikalische Vorgaenge: mit untauglichen Schubladen zur Nulleffizienz. MNU 57 (2004), 115
9. Barke, H.-D.: Das Chemische Dreieck. Unterricht Chemie 13 (2002), 45
10. Johnstone, A.H.: Teaching of chemistry – logical or psychological? CERAPIE 1 (2000), 9
11. Gabel, D.: Improving teaching and learning through chemistry education research: A look to the future. Journal of Chemical Education 76 (1999), 548
12. Ausubel, D.P.: Psychologie des Unterrichts. Weinheim 1974 (Beltz)
13. Piaget, J., Inhelder, B.: Die Entwicklung des raeumlichen Denkens beim Kinde. Stuttgart 1971 (Klett)
14. Pfundt, H.: Urspruengliche Erklaerungen der Schueler für chemische Vorgaenge. MNU 28 (1975), 157
15. Duit, R.: Lernen als Konzeptwechsel im naturwissenschaftlichen Unterricht. Kiel 1996 (IPN)
16. Petermann, K., Friedrich, J., Octken, M.: Das an Schuelervorstellungen orientierte Unterrichtsverfahren. CHEMKON 15 (2008), 110
17. Petermann, K., Oetken, M.: Das an Schuelervorstellungen orientierte Unterrichtsverfahren – exemplarische Vorstellung des Unterrichtsverfahrens an Hand einer Unterrichtseinheit zum Kohlenstoffkreislauf. Vortragsabstract, Wissenschaftsforum Chemie 2007, Universitaet Ulm. Frankfurt 2007 (GDCh)
18. Barke, H.-D., Doerfler, T.: Das an Schuelervorstellungen orientierte Unterrichtsverfahren: das Beispiel Neutralisation. CHEMKON 15 (2008)

19. Barke, H.-D., Strehle, N., Roelleke, R.: Das Ion im Chemieunterricht – noch Vorstellungen von gestern? MNU 60 (2007), 366
20. Sileshi, Barke, H.-D.: Chemistry Misconceptions: Evaluating Remedies Based on Tetrahedral-ZPD Metaphor. Muenster 2007 (Paper presented at the 3rd Symposium of Chemistry Education, University of Muenster)
21. Johnstone, A.H.: Chemistry teaching – Science or alchemy? Journal of Chemical Education 74 (1997), 268
22. Mahaffy, P.: The future shape of chemistry Education. Chemistry Education: Research and Practice 5 (2004), 229
23. Vygotsky, L.S.: Mind and Society: The Development of Higher Mental Processes. Cambridge 1978 (Harvard University Press)
24. Shulman, L.S.: Those who understand: Knowledge growth in teaching. Educational Researcher, 15 (1986), 4

Further Reading

bimbola, I.O.: The problem of terminology in the study of students' conceptions in science. Science Education 72 (1988), 175
Abraham, M.R., Wilkinson, V.M.: A cross-age study of the understanding of five chemistry concepts. Journal of Research in Science Teaching 31 (1994), 147
Boujaude, E.B.: The relationship between students' learning strategies and the change in their misunderstandings during a high school chemistry course. Journal of Research in Science Teaching 29 (1992), 687
Bowen, C.W.: Think-aloud methods in chemistry education. Journal of Chemistry Education 71 (1991), 184
Carr, M.: Model confusion in chemistry. Research in Science Education 14 (1984), 97
Driver, R.: Pupils' alternative frameworks in science. European Journal of Science Education 3 (1981), 93
Fensham, P.J., Garrard, J., West, L.W.: The use of concept cognitive mapping in teaching and learning strategies. Research in Science Education 11 (1981), 121
Garnett, P., Garnett, P., Hackling, M.: Students' alternative conceptions in chemistry: A review of research and implication for teaching and learning. Research in Science Education 25 (1995), 69
Gilbert, J.K., Watts, D.M.: Concepts, misconceptions and alternative conceptions: Changing perspectives in science education. Research in Science Education 10 (1980), 61
Gilbert, J.K., Osborne, R.J., Fensham, P.J.: Children's science and its consequence for teaching. Studies in Science Education 66 (1982), 623
Hihnston, K., Scott, P.: Diagnostic teaching in the classroom: Teaching – learning strategies to promote concept development in understanding about conservation of mass and on dissolving. Research in Science and Technology Education 9 (1991), 193
Kuiper, J.: Student ideas of science concepts: Alternative frameworks? International Journal of Science Education 16 (1994), 279
Lee, O., Eichinger, D., Anderson, C., Berkheimer, G., Blakeslef, T.: Changing middle school students' conceptions of matter and molecules. Journal of Research in Science Teaching 30 (1993), 249
Nakleh, M.B.: Why some students don't learn chemistry. Journal of Chemical Education 69 (1992), 191
Osborne, R., Gilbert, J.: A method for investigating concept understanding in science. European Journal of Science Education 2 (1980), 311
Posner, G.J., Gertzog, W.A.: The clinical interview and the measurement of conceptual change. Science Education 66 (1982), 195

Ross, B., Munby, H.: Concept mapping and misconceptions: A study of high-school students' understanding of acids and bases. International Journal of Science Education 13 (1991), 11

Sutton, C.R.: The learners prior knowledge: A critical review of techniques for probing its organization. European Journal of Science Education 2 (1980), 107

Taber, K.S.: Chemical Misconceptions – Prevention, Diagnosis and Cure. Volumes 1 and 2. London 2002 (Royal Society of Chemistry)

Treagust, D.F.: Development and use of diagnostic tests to evaluate students' misconceptions in science. Journal of Science Education 10 (1988), 159

Wandersee, J.H., Mintzes, J.J., Novak, J.D: Research on alternative conceptions in science. In: Gabel, D.: Handbook of Research in Science Teaching and Learning. New York 1994 (Macmillan)

Watson, R.J., Prieto, T., Dillon, J.S.: Consistency of students' explanations about combustion. Science Education 81 (1997), 425

Fig. 3.1 Concept cartoon concerning the conservation of mass [1]

Chapter 3
Substances and Properties

The main purpose of chemistry lessons is to acquaint students with nature or everyday phenomena and help them to understand what is going on in nature: all guidelines for chemical education, schoolbooks and most curricula should aim at achieving this goal. In addition, the principles of psychology require to offer phenomena which children and students have observed on their own. Based on these observations, they will find initial explanations and will develop their cognitive structure.

However, scientific concepts are often needed to understand the simplest natural phenomena. Whether it is the evaporation of a puddle or the burning of a piece of charcoal, correct interpretation can only be possible by studying the concepts of smallest particles, of the structure of matter, of the formation of atoms, ions or molecules. As long as young people are not aware of these concepts, they tend to develop their own ideas like, "the sun soaks up water from the puddle" or "weight decreases by burning" (see Sect. 2.1). In order to be able to eliminate such erroneous interpretations, chemistry teachers should be aware of preconceptions in the area of

– animistic modes of speech,
– concepts of transformation,
– concepts of miscibility,
– concepts of destruction,
– concepts of combustion, or
– concepts of "gases as non-substances".

If teachers know about most of the preconcepts in these areas, they can better prepare for chemistry lessons; they should be able to discuss these concepts critically with students, and with the help of special media and technology, they will help students to overcome the preconcepts. It is especially useful to have suitable chemical experiments available, which can be performed after such discussions on these preconcepts. In this chapter, suitable methods of teaching will be presented *without using the particle concept*. In Chap. 4, a follow-up will be offered with diagnostic tools and teaching suggestions concerning the particle concept and Dalton's atomic model.

3.1 Animistic Modes of Speech

The mental development of beginners in chemistry can be associated primarily with concrete thought operations according to Piaget: they are mainly fixed in their ideas on the specific object. This leads to the fact that they tend to describe such phenomena in a concrete-pictorial manner and in a magic-animistic mode of speech. The following examples have often been experienced:

- wood *will* not burn, the flame *will* extinguish, and the flame *devours* the candle,
- acids *attack*, they *eat* base metals, rust *guzzles* the body of the car, etc.

Pupils' explanations often correlate with simple analogies, causes are personified:

- sodium metal reacts with water "like a fizzy tablet",
- when copper sulfate dissolves, it is like "the way red cabbage runs in water",
- crops grow in fields *so that* people can eat,
- wood burns, *so that* one can warm oneself, etc.

Püttschneider and Lück [2] discuss the role of animism in the teaching of chemical topics and propose the following points: "(1) Using animisms consciously generates a positive association to the subject of chemistry, (2) animisms tend to have a lasting motivational effect because they help the students comprehend the subject, (3) the students are aware of the model character of the animisms" [2]. Initial results show that "animisms lead to better understanding thereby having a lasting motivational effect" [2].

Teaching and Learning Suggestions. At first, it is obvious that the above pupils' statements use everyday language and the pertinent observations based on this fact are easy to comprehend: every person knows what is meant when somebody says, "the wood will not burn". Therefore, one should not state these forms of expression as incorrect.

If there is enough lecture time and the question comes up from the students regarding a so-called "will of wood", one should elucidate this topic. One could discuss whether a damp piece of wood is always hard to ignite, or if wood contains a "will" of its own, if it is "sometimes easy to light and sometimes difficult, depending on the wood's will". The question arises as to whether it could be, that small dry pieces of wood are always easy to ignite and continue to burn? In this discussion, the statement "the wood will not burn" can be restated depending on the conditions under which wood can easily be ignited or not. One could even show that different types of wood react differently and that, for instance, birch bark belongs to the type of wood, which contains easily ignitable resin and is therefore ideal for starting up a campfire. When the fire is finally lit, one can warm oneself from it: man uses fire with the intention of having "light and warmth". Wood burns when a person wants it to and when he or she is in a position to ignite it – it is not the wood "which decides" when and if a person needs warmth.

The teacher has to decide if his or her pupils like to discuss such philosophical ideas – if not, he or she has to stop it or postpone it for a later consideration.

3.2 Concepts of Transformation

In lectures, characteristic changes of substances are generally described as chemical reactions and the formation of new substances with new properties. Nevertheless, students do not usually associate the appearance of new matter with chemical reactions; rather one likes to imagine the "same substance" with new properties:

- copper roofs *turn* green,
- silver metal *turns* black,
- iron *changes* to a rusty color,
- or, water *becomes* red.

There seems to be a "carrier of properties" in the minds of students which "somehow remains intact" in matter and which constantly changes its appearance: copper metal can appear to be red-brown colored as well as green; iron exists as light grey as well as rusty brown. Also many ancient philosophers believed the existence of "basic matter" which can change properties – and even up to about 200 years ago, alchemists tried to find an "elixir" which could transform mercury or lead into gold. Such thinking still exists amongst our young students.

Young people are unlikely to philosophize about "original matter" as the Greeks did but, with the influence of their everyday language, will easily come up with a "transformation concept". Boiling an egg produces a soft or hard egg. The longer you cook it, the harder it will be: the egg remains. A steak can be cooked "rare", "medium", or "well-done" – one can transform it through frying, it "changes" through the treatment but the steak remains. As long as such everyday language continues to have such an influence on young people, one cannot ignore the development of a transformation concept.

In his empirical research Schlöpke [3] finds "statements which allow for a substrate concept". After heating up red cobalt chloride and observing the change to blue color, "one is convinced that after the heating process the substance remains. In order to explain the change in appearance, two different points of view exist: One group of students think that the substance is composed of layers whereas the color is probably attached to a material carrier (this does not need to be identical to the cobalt chloride). People who follow this proposal use the terms layer, color layer, wrapper, evaporated color, scorched, burnt (in terms of being destroyed). They also visualize that the color has perhaps retreated into the core, has gone back inside. The other group constructs a set of hierarchical colors: first red, then blue, black, white and transparent. However, a loss in properties does not change the opinion of students that the same original matter stays, only having been reduced in properties" [3].

Teaching and Learning Suggestions. It is quite a difficult job to explain which new matter arises during certain chemical processes in the kitchen – like cooking, frying and baking. It is much easier to take usual examples of the laboratory to explain chemical reactions, to visualize with models for understanding and to formulate chemical equations.

As soon as one comes across a "green copper roof", one should attempt to remove the green layer until one reaches the pure red-brown copper metal. If one dissolves the green substance in diluted hydrochloric acid, bubbles form. The addition of limewater helps to prove the existence of carbon dioxide. The green substance must be a type of copper carbonate, a completely different substance than the red-brown metal. The formation of green carbonate can be explained by reactions of copper with the solution of carbon dioxide in rainwater, or the formation of blue copper sulfate by the reaction of copper with industrial "acidic rain".

Different properties have to be associated with different substances. Comparing the green substance of the copper roof with various copper carbonates, copper sulfates and copper chlorides, one realizes that there are different green or blue copper compounds, but no green copper exists! The discussion should result in the statement that copper is covered with a green copper compound, called copper hydroxo carbonate.

Should one not have access to a roof covered with copper and green copper carbonate, then, it is possible to carry out a model experiment (see E3.1). A copper sheet is folded in the shape of a letter envelope and intensely heated on the outside with the burner. A blackening on the outside is observed. However, upon unfolding it, there is no color change on the inside. A second experiment brings further proof (see E3.2). One places a small sample of copper (sheet or wool) in a test tube with a side tube and a stopcock, evacuates it with a water aspirator (or a vacuum pump) and heats it: no change occurs with the metal. If one opens the stopcock, the hot metal instantly "turns black". Analysis of both experiments leads to the conclusion that air or oxygen of the air is responsible for changes of metals: copper is and remains red-brown, it reacts with oxygen of the air to form black copper oxide. It is correct to say "the substance turns black" when describing this experiment, when writing laboratory reports. In the interpretation of this observation, it should however be noted that a new substances is formed, black copper oxide:

$$\text{copper(s, red-brown)} + \text{oxygen(g, colorless)} \rightarrow \text{copper oxide(s, black)}$$

In the case of "silver turning black", experiments can be used to show reactions of some metals with yellow powder of sulfur to form black sulfides: copper sulfide and silver sulfide. One finds the same black silver sulfide, which is caused by the reaction between silver cutlery like knives, spoons and forks with hydrogen sulfide from the air: one can remove the black silver sulfide from the surface of pure silver.

It is possible to recover the silver by heating the black silver sulfide (see E3.3). Similarly, copper can also be recovered from copper oxide through the reaction with hydrogen (see E3.4). However, this leads to the question of how the elements are "stored" in metal compounds (see Sect. 2.2). This question is difficult and only answered at a level of Dalton's atomic model: metal atoms and non-metal atoms remain during these reactions.

The fact that pulverized metal crystals generally turn black should perhaps be shared with students. An electrolysis of a silver nitrate solution with DC voltage leads to a deposition of black metal lumps. The metal forms silver crystals only when the concentration of free silver ions is strongly reduced through stable complexes, i.e. when the silver nitrate solution is mixed with sodium cyanide solution or with ammonia solution.

The third example is that of iron with "different forms of appearance". It is possible to find a thoroughly rusted object on which the rust has taken such a hold that it begins to flake off. By scraping off the rust and pulverizing it with the help of a mortar and pestle, one obtains a powder, which can be compared with the iron oxide in store of chemicals. The red-brown powder is a different substance than iron – it is formed through the rusting process, i.e. through the reaction of iron with oxygen and water. Rust is in fact, from a chemical point of view, a mixture of iron oxides and iron hydroxides.

In the fourth example, water "becomes red" by dissolving cobalt chloride or adding a special acid–base indicator. In this case, we also have to deduce that a new substance is formed. It is worth discussing whether this substance should be called a "red-colored solution" or a "red solution": what is coloring what? Water is a colorless liquid at all times. If water turns red by adding cobalt chloride or indicator solution, a chemical reaction occurs, the new product is a red colored solution: no pure water remains!

The drink commonly known as "Cherry Water" contains water, but also up to 40% volume of ethanol besides the cherry juices. One really cannot call this solution "water". Dissolving sugar in water leads to a colorless sugar solution and one should not call it "sweet water". In chemistry laboratories, we call some detection reagents "limewater" and "baryta water": one should better use more accurate and meaningful labels like "calcium hydroxide solution" or "barium hydroxide solution"!

3.3 Concepts of Miscibility for Compounds

The concept of mixture and separation also plays a large role for students in the interpretation of chemical reactions and the composition of new substances:

- copper sulfide *contains* copper and sulfur,
- water is *composed* of hydrogen and oxygen,
- hydrocarbons *have* carbon and hydrogen.

To begin, the statement that "one water molecule is composed of two H atoms and one O atom" is correct. The terms "contain", "made up of" or "composed of" are valid at this level of Dalton's atomic model. However, how would these words relate when one is talking about substances on the macroscopic level?

The above statements suggest a concept of mixed substances in compounds. Greek philosophers have already discussed "mixing of elements like air, water, earth and fire" while observing the enormous diversity of matter: with the ideas of elements and their mixtures they wanted to give an explanation. Naturally, mixtures do exist between various substances, e.g. all the cleaning substances in the bathroom or many substances in the kitchen. In addition, metals and sulfur can be mixed: one can see the individual components, e.g. red-brown copper crystals and yellow sulfur crystals in a mixture of both. However, when heated, both chemicals combine to form a black homogenous substance, which points neither to copper nor to sulfur. Because this new substance results from a chemical reaction of copper and sulfur, it is commonly known as copper sulfide. Only an expert is aware that copper sulfide crystals contain special amounts and arrangements of copper ions and sulfide ions. The beginner may think of a mixture of copper and sulfur, but he or she cannot have appropriate ideas of how copper and sulfur are arranged in copper sulfide – this will still be a puzzle till information about atoms and ions are presented in lectures.

Teaching and Learning Suggestions. First, we should show students examples of *heterogeneous* mixtures: granite rock, little crystals of three different minerals mixed to a barren rock, salt and sand mixtures, etc. The sample could also be a laboratory mix of copper crystals and sulfur powder: one can observe two different types of crystals with a magnifying glass; one can even separate them with a tweezers. One can compare this mixture to pure copper sulfide produced from the chemical reaction of copper and sulfur. If one sees only one single substance type, as in the case of copper sulfide, then one has a *homogeneous* pure substance.

The common argument that only elements are pure substances and that "sodium chloride contains sodium and chlorine" should be rejected. This is because one measures constant values for the density or for the melting point of sodium chloride, and will find such constants of other compounds in well-known published tables. Having a defined density, melting and boiling point, the related substance is considered as a pure substance, e.g. water or ethanol. Pure substances are elements *and* compounds!

One can still discuss statements like "sodium chloride is a compound of the elements sodium and chlorine" or "hydrogen and oxygen are combined in water" – however, they do not give any explanation to the presence of the elements in the compound. One should avoid statements like "water contains hydrogen and oxygen" – only at the level of the Dalton atomic model chees it becomes clear that "a water molecule is composed of two hydrogen atoms and one oxygen atom".

Fig. 3.2 Concept cartoon concerning composition of vapor [1]

3.4 Concepts of Destruction

Through the fascinating in the observation of flames and the visible destruction of matter in camp fires, house or forest fires, students are almost forcibly led to a "destruction concept", mainly because of the everyday language:

- candles, spirits or gasoline completely "burn away", carbon "smolders", wood is "charred",
- plants "rot", animal carcasses "decay", food is "digested",
- water "evaporates", stone is "weathered", sandstone "crumbles",
- metal "corrodes", iron "rusts",
- fat spots are "removed", remnants are "destroyed", etc.

Barker [4] mentions other concepts of destruction as described by many young people through the "optical destruction" of substances:

- Substances can cease to exist, still there remain taste and smell: sugar is destroyed by dissolving in water, the sweet taste remains ("sweet water"), drops of liquid perfume disappear with time but the scent remains.
- The mass of portions of material is not a decisive characteristic, the existence of mass-less material is generally accepted by young students without any reservations: particularly vapor of volatile fluids like gasoline or alcohol "lose their mass upon evaporating".
- Substances that exist can "effortlessly disappear": should the water level in a tank on a hot summer's day be lowered, young people do not find it necessary to explain, the water just disappeared, the water is no longer there!

Students do not generally have a scientific view of the concept of preservation of mass and energy. Thus, it is not possible for them to reflect properly on these phenomena without expert assistance. They do not really stand a chance of coming up with or upholding their own concept of destruction or arriving at an adequate preservation concept based on their own observations. The strong influence of the language which surrounds them everyday, even many media advertisements hinder scientific ideas. The purpose of lectures about conservation of mass and energy is in this sense very important because the acquired preconcepts are very established: they must be discussed in depth before students will accept the idea of the preservation of mass.

Teaching and Learning Suggestions. It is necessary at first to discuss and reflect the statements, i.e. that "when water evaporates, it no longer exists". Students know about the existence of damp and dry air, of cloud formation, and how rain comes about when the humidity level in the air is too high. If one takes into account the humidity of the air when discussing evaporation of water, then one can deduce that water does not "disappear", but it vaporizes and remains mixed with air as invisible steam. It is also necessary to reflect that the evaporation of water does not only occurs at the boiling point of 100°C (at normal pressure), but also under this boiling temperature: at room temperature water of rain puddles evaporates (see Sect. 2.1). It is necessary to compare the process of evaporating and boiling, and to state that the general term "vaporization" includes both. For further explanations the particle model of matter would be advantageous.

It is particularly useful to use experiments to show that there is no "destruction" of matter during the process of evaporation, but rather that matter remains. It is possible by using a very sensitive scales to observe the evaporating process of acetone; one can observe the decreasing mass, but in the second step one can see however, that a large volume of gaseous acetone vapor is formed (see E3.5). If one fills gaseous butane in a special liquid-gas syringe and presses the gas with the piston, a part of butane changes to liquid and remains liquid

3.4 Concepts of Destruction

under pressure (see E3.6). If one allows liquid butane to evaporate in the syringe the gas results again, if one applies pressure to the portion of gas, then one will end up with the same amount of liquid butane as before (see E3.6). Also with this experiment the students can observe, that a gas is a substance which does not disappear; which does not go "away" after evaporation.

Regarding the well-known "disappearance" of lithium or sodium in reaction with water, it must be demonstrated that the gas hydrogen is formed and new white substances – hydroxides of the metals, appear after the water has been evaporated from the alkaline solutions (see E3.7). So the metals are reacting

Fig. 3.3 Concept cartoon concerning destruction of matter by burning [1]

with water to form new substances, weighing all reactants and products the masses will be the same, the law of conservation of mass can be applied.

The wrong idea of "removing and destroying" a greasy stain from clothes by using a cloth dipped in an organic solvent (gasoline, alcohol, etc.) may be interpreted by dissolving a small amount of fat in gasoline and weighing the solution (see E3.8): the masses of fat and gasoline before and after dissolving are the same. Dropping some drops on a filter paper, the fat still remains after the evaporation of the gasoline (see E3.8).

3.5 Concepts of Combustion

Concepts of a carrier of properties ("copper turns black by heating with a flame") as well as concepts of destruction ("wood is burned") go back to students' preconcepts of burning [5]. After several years observing the ever-fascinating burning process, i.e. of paper, wood, charcoal, spirits or gas, students are lead to the statement that "*something* is going up in the air, that *something* disappears and only a few ashes remain". This *something* as described by our young students can be compared to the "Phlogiston" of the ancient scientists: the word was phrased by the German chemist Stahl in the 17th century (see Sect. 1.3).

Students were asked what actually happens to the particles of magnesium during the well-known experiment of burning the metal in air. Most students noted one of the correct reaction symbols like "Mg + O → MgO". But they argued that a part of the magnesium disappears into the air and another part remains in the form of white ashes. One student writes: "magnesium is composed of two kinds of particles; one evaporates in the burning process, the other remains as magnesium oxide" [6]. This student even draw a nice figure fitting to his description (see Fig. 3.4).

An additional diagram (see Fig. 3.5) underlines what other researchers have come across. They shared similar ideas by their students regarding the burning process of iron wool and of phosphorus [7]. Astoundingly enough, even older students who have had many years of chemistry do not easily let go of this destruction concept. One 10th grader student claimed that, "according to the formula it must be possible to get carbon from CO_2, but it is impossible to get a black substance out of a colorless gas" [5].

Teaching and Learning Suggestions. It is certainly advisable to discuss and compare preconcepts of burning processes with the students. Doing so, one can draw their attention to the reaction of substances with air or with oxygen. Here, one can learn through fire fighting: fires can generally be extinguished through cooling down the burning matter with water or through preventing air from getting through or by using sand, blankets or carbon dioxide gas. At the end, students will learn that the burning process is a reaction that involves oxygen of the air.

A variety of experiments is particularly good in helping to get the point across. Demonstrations or experiments are done that compare burning items in air and in pure oxygen: the burning occurs much more intensely in pure oxygen than in air. In

3.5 Concepts of Combustion

> **Question:**
>
> You have learned that when magnesium is burned a white powder forms.
> 1. What is the chemical equation for the burning of magnesium?
>
> **Answer:** $2Mg + O_2 \longrightarrow 2MgO$
>
> 2. Write down and sketch your idea for what happens to the magnesium particles during the magnesium burning process.
>
> **Answer:** Magnesium consists of two kinds of particles, one is vaporized by the burning process, and the other remains as magnesium oxide.
>
> **Drawing:**
>
>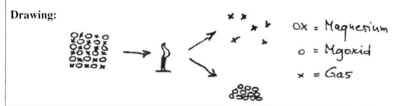

Fig. 3.4 Image of a 16-year old student regarding combustion of magnesium [6]

addition, air can be shown to be a mixture of oxygen and nitrogen. One could also discuss the observation of burning candles and their extinction under a beaker as a possible reason for the presence of reduced oxygen level in the air. Further experiments would also help to understand the role of oxygen (see E3.1 and E3.2): black copper oxide is formed through the reaction of copper with oxygen in the air.

In other experiments of burning processes, one should use good scales. Historically, the scales have played an important role regarding knowledge of the oxidation theory: French scientist Lavoisier observed the formation and

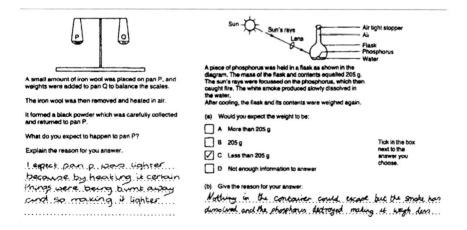

Fig. 3.5 Statements regarding the burning process of steel wool and phosphorus [7]

decomposition of mercury oxide in the middle of the 18th century; his measurements helped to overturn the Phlogiston theory (see Sect. 1.3).

Experiments of burning metals could show that the mass increases in an *open container*: steel wool becomes heavier upon smoldering; the mass of the combustion product of magnesium is larger than it was at the start of the experiment (see E3.9). Because many of these experiences are new for most students, a classical cognitive conflict regarding common experience of "becoming lighter" can be developed. The increase in mass is convincingly attributed to the amount of oxygen in the air, which reacts with the metal: gaseous oxygen is fixed in solid metal oxides.

Fig. 3.6 Concept cartoon concerning conservation of matter [1]

3.5 Concepts of Combustion

It must be pointed out, that the mass neither increases nor decreases when a metal burns; mass is rather conserved: If one weighs substances before and after a reaction with oxygen in a *closed container*, it is possible to demonstrate experimentally the conservation of mass. This can be done by suspending steel wool in a closed test tube (see E3.9).

A more serious problem for young students is to recognize the development of gaseous and colorless oxides in the combustion of charcoal, wood or candle. The experiences of students in observing burning charcoal are that "charcoal is gone", only white ashes remain. To get the scientific idea about the combustion of charcoal, little pieces of coal are placed in a big round flask filled with oxygen and closed by a rubber balloon (see E3.10). The mass of the flask is measured carefully. The coal pieces are heated in the flask until they burn and "disappear" by shaking the flask. Weighing again, the "empty flask" shows the same mass as before, adding limewater the gas carbon dioxide is proved. This experiment convinces students to accept that charcoal is not "destroyed" by combustion, but has reacted with oxygen to form carbon dioxide:

carbon(s, black) + oxygen(g, colorless) → carbon dioxide(g, colorless)

Students also have the mental model that candles "disappear" by burning. So first it has to be shown that the flame of a candle dies out as soon as a beaker is placed over it: an obvious lack of adequate amounts of air can be supposed. Not so apparent is the production of carbon dioxide and steam of water by combustion because these gases are colorless and invisible. More convincing is the water layer which can easily be seen when a cold beaker is placed upside-down over a burning candle (see E3.11).

One should be careful in interpreting the results of the following experiments. When a glass cylinder is placed over a floating tea light, as expected, the flame dies out, and the water level rises quite a bit in the cylinder (see E3.11). One cannot, however, talk about "used" oxygen or about the disappearance of "20% oxygen in the air", It is more appropriate to talk about the formed colorless gas carbon dioxide. Because steam is also formed by the burning candle and condenses when the flame is extinguished, the volume in the cylinder decreases. Also hot air cools down when the flame is extinguished and causes the volume to decrease and the water level inside to rise. There are other more convincing experiments which can determine the amount of oxygen in the air by forming solid oxides of copper, iron and phosphorus (see Sect. 3.6).

The formation of steam and carbon dioxide can be convincingly demonstrated using a beaker to see condensed water and to prove the appearance of carbon dioxide through testing with limewater (see E3.10). To demonstrate the increase of mass like in similar experiments of burning metals to form solid metal oxides, one has to absorb the invisible gases. To show this effect with a candle, one has to light a candle on the scales in such a way that the gaseous products steam and carbon dioxide are caught by a glass cylinder and absorbed by special chemicals like sodium hydroxide: while the candle burns its mass increases (see E3.11).

In order to disqualify the statement that "there is no solid carbon obtained from gaseous carbon dioxide", the reaction of colorless carbon dioxide gas with burning magnesium is shown (see E3.12): In addition to the white magnesium oxide, some black soot is formed on the inner wall of the cylinder, which can be retrieved with the finger: carbon. These experiments also will show that gases are substances with specific properties (see Sect. 3.6).

3.6 Concepts of "Gases as not Substances"

Many experts of centuries ago did not recognize air as a substance nor did they know the different colorless gases present in the air. It is just as difficult for today's children. Because humans are constantly surrounded by "weightless air" and because it is common knowledge that warm air rises, children do not imagine that portions of air have any mass therefore air is not seen as a substance. Münch [8] could therefore show in empirical studies that approximately 50% of students between 10 and 16 years believe that a ball, which is fully inflated with an ordinary air pump, should be lighter than a non-inflated one [8].

On the topic of gases, many incorrect ideas circulate in everyday language. Weerda [9] collected the following statements from children at school:

- Fresh air is "good" air, there also exists "bad" air, air without oxygen is "bad",
- a chimney needs "supply air" and delivers "exhaust air",
- cars emit "exhaust" to the air,
- colorless gases are "air" or similar to air, water evaporates "and forms air",
- gases are combustible, they are there for cooking and heating,
- gases are dangerous, are explosive, are poisonous, and
- gases are "liquids", one finds "liquid gas" in lighters [9].

Voss [10] also finds similar associations in his questionnaires to students of grades 7–10 regarding the following question: "What do you think gas is"?

- Gases are invisible, one cannot taste them,
- one needs gas in order to make a lighter work,
- gas is a good burning agent, it is used in chemistry to heat up substances,
- when fire touches gas, then the container with gas explodes,
- one always breathes in a little gas, but gas is not particularly healthy,
- gases are generally more dangerous than solid matter,
- gas is invisible and is contained in the air,
- gas is liquid, liquid gas is in lighters, gas can appear liquid,
- there is heavy and light gas: carbon dioxide and helium,
- gas is lighter than a solid substance, it can infiltrate anywhere [10].

Although the kids had heard quite a lot about gases from their teachers and had experimented with them, and although the three physical states are known, "experienced properties of gases are hardly internalized, and kids tend to hold

3.6 Concepts of "Gases as not Substances"

on to their preconcepts for as long as possible and can hardly be convinced by scientific points of view" [10].

It is therefore necessary to clearly identify air as a substance, better as a mixture of different gases and to distinguish air from other colorless gases, which are not part of the air: they also play an important role in everyday life. So it is necessary to have suitable experiments that demonstrate all these facts, better students can carry them out themselves.

Teaching and Learning Suggestions. Regarding the question of whether gases are substances or not, it is possible to show that gaseous butane changes to liquid butane under pressure (see E3.6). To demonstrate that one can even obtain liquid air or liquid nitrogen, it is particularly spectacular when one places some liquid nitrogen in a Dewar flask and measures its boiling temperature of about minus 196°C. One can also produce blue-colored liquid oxygen through condensation of oxygen by using a test tube and cooling oxygen down with the help of liquid nitrogen. In a follow-up discussion, one could point out that temperature and pressure are key in whether or not a substance is solid, liquid or gaseous – air is known to be a mixture of gases under normal conditions. Because it may not always be easy to obtain liquid nitrogen or liquid air, at this time, no experiments will be mentioned here – the literature should be helpful in finding suitable activities.

Another way to convince young students that gases are substances, their density should be measured (see E3.13): Evacuate a round flask with stopcock with the help of a water aspirator or a vacuum pump, exactly weigh the flask on an analytical scale and, after filling up with 100 ml of air from a syringe the flask is weighed once again: a net mass of 0.13 g is registered (see E3.13), the density of 1.3 g/l is calculated. If there is no vacuum source available, one can put 100 ml of air or another gas in addition to the air in the flask: the compressed gas can be weighed. The density of additional gases can also be established this way, i.e. carbon dioxide, butane, etc. Differentiating their densities from that of air will help to understand the characteristics of gases, will help to look at gases as substances. It is especially important to point out that, due to their low values, one should write the density of gases in g/l, rather than in g/ml.

One could also demonstrate the well-known gases, like oxygen, nitrogen and carbon dioxide, in addition to combustible gases like hydrogen and butane, showing their differences with the use of a burning or glowing wooden splint (see E3.14). Because both nitrogen and carbon dioxide gases do not support combustion, the question arises as to how to differentiate them. Both the test with calcium hydroxide solution (lime-water test) and "pouring" carbon dioxide onto a burning tea-light in a beaker can show the differences of nitrogen and carbon dioxide (see E3.14): nitrogen does not react with limewater nor is it heavy enough to extinguish a candle in the beaker.

Along the same line, the amount of oxygen in air (about 20% by volume) can be demonstrated with the reaction of steel wool or phosphorus in a closed container (see E3.15): both substances produce solid oxides, which remove oxygen quantitatively out of the air.

Finally, as clarification of the term "liquid gas", the condensation of butane can be repeated as before (see E3.6). The result is a large drop of liquid butane in presence of gaseous butane. It is important to point out that – like in the butane pump – both liquid and gaseous butane can be found together in lighters and camping gas cartridges, the commonly called "liquid gas" does not exist. It is also possible to interpret these phenomena by equilibria between liquid and gas.

Conclusion. Students' alternative concepts should be discussed carefully and compared in detail; students should be encouraged to develop their own individual interpretations by observing and investigating for themselves. Should this lead to the introduction of appropriate concepts, then the teachers could convincingly support them or, better still, let the students carry out the appropriate experiments themselves. In other cases it can be advantageous to teach first the scientific concept and to discuss the misconceptions later (see Sect. 2.3). In every case teachers should not continue to teach "by the book" instead they should integrate preconceptions and misconceptions and reflect on the teaching-learning process. Students will not be able to fully understand and may return to their original ideas – but they start to reflect on their own preconcepts!

If stimulating activities and experiments are demonstrated or carried out by the students themselves, their newly-acquired knowledge will make more sense than their original ideas – they will develop a sustainable scientific concept. In evaluating new information or experiments, one could continually present arguments, which will help cement the scientific concept in their minds. After some time this will lead to the acceptance of the new concept, enabling them to become embedded in their cognitive structure. Furthermore, it will, in the sense of constructiveness, allow for a new individual with an integral cognitive structure to get rid of separate compartments for old and new ideas (see Sect. 2.3).

3.7 Experiments on Substances and Their Properties

E3.1 Heating a Copper Envelope

Problem: This experiment demonstrates that copper does not "change" color from red-brown to black, but rather that a black substance always develops when air is present: copper reacts with oxygen in the air, forming black copper oxide. If no air is available in the inside of the closed metal envelope, then the red-brown copper remains nearly unchanged.

Material: Crucible tongs, burner; sheet of copper metal.

Procedure: Fold a sheet of copper metal in half. Fold again and press together with crucible tongs the open sides of the sheet. Heat this envelop in a hot burner flame for 10–20 s. Cool the envelop and unfold it. Compare the inner and outer parts.

Observation: The outer side of the red-brown metal envelope turns black upon heating: copper oxide. The black substance separates and falls off upon unfolding the envelope, on the inside of the envelope only red-brown copper is to be seen, no black substance.

Disposal: The copper sheet can be used for further experiments following the complete removal of the black copper oxide.

E3.2 Heating of Copper in Vacuum and in Air

Problem: In order to discuss the reaction with air as an important component, one could heat the copper without air in an evacuated test tube: copper remains unchanged after heating in the vacuum. Immediately after this, air is introduced by opening the test tube: in the same moment hot copper turns to black copper oxide. With this experiment students should be convinced that the air is responsible for forming the new substance by a chemical reaction.

Material: Vacuum source (water aspirator or vacuum pump), Pyrex test tube with side tube and stopcock, stopper, burner; copper wool.

Procedure: Fill a test tube to one-third with copper wool, close and evacuate. Intensely heat the copper for 20–30 s and open the stopcock.

Observation: At first, there is no change by heating the copper. As soon as the air hits the hot metal, a black substance is formed: copper oxide.

E3.3 Decomposition of Silver Sulfide or Silver Oxide

Problem: Silver is an element and therefore it is not possible to decompose it into different substances. It should not be possible for "black silver", if one calls it an element, to decompose. If, however, one takes the black substance, which we find on silver cutlery, and intensely heats it, a silvery metal and a pungent smelling gas are produced: silver and sulfur dioxide. This experiment shows also that the black substance is a compound: silver sulfide. A similar experiment could be performed with black silver oxide, it decomposes to form silver and oxygen.

Material: Pyrex test tube, burner; silver sulfide or silver oxide.

Procedure: Put a small amount of silver sulfide or silver oxide into a test tube. Heat the test tube with the hot flame of the burner until the black substance is completely decomposed.

Observation: A shiny ball of metal is produced from the black powder silver.

E3.4 Reaction of Copper Oxide with Hydrogen

Problem: This experiment demonstrates that it is possible to "regain" the red-brown copper from the black copper oxide with the help of hydrogen gas: by this reaction copper and water are formed. It is difficult, however, to interpret this reaction on the substance level: one cannot really state "copper oxide contains copper and releases it" because one cannot find red-brown copper in the black copper oxide! On the level of redox reactions and electron transfer it is later possible to give the scientific explanation (see Chap. 8).

Material: Combustion pipe with stopper and exhaust pipe, porcelain boat, test tubes; black copper oxide (wire form), hydrogen gas.

Procedure: Introduce hydrogen gas into the combustion pipe, which contains copper oxide filled in a porcelain boat. If the oxyhydrogen test turns out to be negative, light the hydrogen on the exhaust pipe and heat the copper oxide with a burner. As soon as the reaction sets in, indicated by the formation of metallic copper, remove the burner. Cool down the combustion pipe under a continuous hydrogen stream before the hydrogen supply is turned off.

Observation: Shiny red-brown metal is formed through the exothermic reaction: copper. One can see clear drops of a liquid in the exhaust pipe: water.

Disposal: The copper can be used for further experiments or it can be oxidized in a stream of air, enabling the copper oxide to be used again for the same experiment.

E3.5 Evaporation of Acetone

Problem: Students often describe evaporation of a liquid as the "disappearance" of the substance: "rain puddles *disappear*". In order to show that a liquid is transformed to a gas, one should use a glass syringe and an analytical scales; to show the vapor of acetone which results from the evaporation process. One can also smell the mixture of acetone vapor and air.

Material: Analytical balance, watch glass, Erlenmeyer flask with glass beads and stopper, glass syringe; acetone.

Procedure: Place a few drops of acetone on a watch glass and weigh, after some time weigh it again. In the Erlenmeyer flask connected to a syringe, add a few drops of acetone, warm with hands and shake the glass beads. Open the flask and smell.

Observation: The analytical balance shows a decreasing weight reading until finally all liquid acetone evaporates. In the second step the syringe fills up with a colorless vapor, until the liquid is completely evaporated and the gas weight remains constant. Opening the flask the gas can be smelled: acetone respectively the mixture of air and acetone.

E3.6 Condensation of Butane Gas Under Pressure

Problem: Students are probably aware of butane lighters and call the butane "liquid gas" – although they have observed the liquid butane and the gaseous

3.7 Experiments on Substances and Their Properties

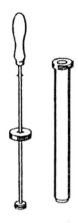

butane in a see-through lighter. Despite such observations, the term "liquid gas" still tends to stick and should be studied in the following experiment. It also shows the characteristics of butane: it can be condensed by using pressure, thereby transforming to a visible liquid with specific boiling temperature of about 0°C.

Material: Gas liquefying pump (see picture); butane (from the cartridge of a butane burner).

Procedure: Open and fill a gas liquefying pump with butane from the cartridge through replacement of the air. Attach the piston, press strongly into the pressure-resistant test tube and lock it in place. Release the lock again and observe the piston. If one touches the tube with evaporating liquid butane, one can feel a cooling effect. Repeat this procedure several times.

Observation: A large drop of liquid is formed during the compression of the gas: liquid butane. When the lock is released, the piston moves out of the tube and the liquid drop turns completely into gaseous butane, the same gas volume as before is to be seen. During evaporation is observed a cooling effect, the temperature of the liquid butane decreases.

E3.7 Dissolution of Metals

Problem: The popular reaction of sodium or lithium with water often leads students to make the statement that the metal "disappears" – like a "fizzing antacid tablet that dissolves in water". On the one hand, one can see gas as a product and the color change by using an indicator solution. On the other hand, after heating the solution and evaporating the water – it is possible to obtain a white solid, sodium hydroxide. Instead of "disappearing", the metals react with water to form a hydroxide solution and the gas hydrogen.

Material: Large glass bowl, cylinder, glass plate, test tubes, beaker, forceps, knife, filter paper; sodium, lithium, phenolphthalein solution, ethanol.

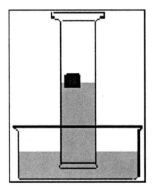

Procedure: (a) Place a glass bowl half filled with water on an overhead projector. Place a piece of sodium on the surface of the water and observe the path of the metal. Repeat the experiment several times. (b) Repeat the previous experiment using lithium. Place a piece of lithium on the surface of the water and observe. (c) Fill water in an upright cylinder and cover it with a glass plate, turn upside down, place under water and uncover it. Put a piece of lithium in that cylinder (Caution: do not use sodium for this experiment). After gas has developed, place the cylinder upright again and light the gas. (d) Take a sample of the solution and test it with indicator solution. Take another sample of the solution, place it in a beaker and boil until dry.

Observation: One can clearly see streaks from the projection of the reaction of sodium metal pieces with water, hear the hissing sound that appears and note the fizzing of the gas produced.

The colorless gas from the reaction of lithium in the cylinder burns in air with a red-colored flame: hydrogen. The solution turns the colorless indicator deep red. By boiling the solution a solid white substance remains in the beaker: sodium hydroxide or lithium hydroxide.

Disposal: The solutions are diluted sufficiently so that they can be disposed of down the drain. Remnants of sodium or lithium can be dissolved in ethanol and disposed of down the drain.

E3.8 Dissolution of Grease

Problem: When "removing" grease stains in clothing, the grease "disappears" or "goes away", in the minds of children. In order to connect the removal of grease to the dissolution process of grease in gasoline, one should clearly indicate that the pure gasoline evaporates leaving no residue. However, if one evaporates a greasy solution, one will be left with a grease stain. For showing the conservation of mass by producing the solution, gasoline and oil are weighed first, after dissolving the solution will show the same mass as both portions of substances before.

Material: Test tubes, scales; gasoline, olive oil, solution of olive oil in gasoline (greasy solution), filter paper.

Procedure: (a) First weigh some drops of oil and a few ml of gasoline. Dissolve the oil in the portion of gasoline and weigh again. (b) Place a few drops of gasoline on a filter paper and at the same time a few drops of the greasy solution on another paper. Observe both papers carefully.

Observation: (a) The masses remain the same before and after dissolving. (b) The stain from the pure gasoline becomes smaller and smaller until it can no longer be seen. The other stain also gets smaller but a clear grease stain remains on the paper.

E3.9 Burning Metals on a Balance

Problem: Because of everyday experiences, children tend to believe in the "loss of mass" or of the material "becoming lighter" when alcohol, paper or candles are burned. Even after observing that iron wool glows, students state that the formed black substance "is lighter than the iron before". Using a balance, one demonstrates that the black substance solid iron oxide is heavier due to the reaction with oxygen. These experiments show the increase in mass accompanied by the formation of a solid substance, like metal oxides or phosphorus oxide in an open system.

If gases are formed as in the case of the oxidation of carbon or sulfur, then special substances must first absorb the gaseous reaction products before it is possible to weigh these products (see E3.10). In order to include the conservation of mass during reactions to substances "becoming heavier", one should again repeat metal reactions in a closed container and compare results: if metal sample and oxygen are weighed together, the mass of reaction products do not change by the reaction using a *closed* flask, in contrast to *open* systems before.

Material: Beam balance, digital balance, crucibles with lids, test tubes, balloons; steel wool, magnesium ribbon.

Procedure: (a) Light a piece of steel wool which is hanging on the side of a leveled beam balance; if necessary lightly blow on the steel wool in order to speed up the reaction and to see the burning process more clearly. Observe the color of the product. (b) Weigh a rolled 10 cm magnesium ribbon in a pre-weighed crucible. Then heat the crucible in an intense hot flame. Cover the crucible slightly, the magnesium will burn. Again weigh the cooled-down crucible. (c) Fill a test tube to about half with steel wool, fit with a balloon and weigh. Heat the test tube until the wool reacts, let the contents of the test tube cool in the air and weigh again.

Observation: (a) The side of the beam balance containing the burning steel wool is lowered; the shiny metal changes into a black substance. (b) A white combustion product is formed from the brightly glowing magnesium and this smoke caught by the crucible; the balance shows a larger mass than before.

(c) The test tube heated with iron wool fitted with a balloon weighs the same before and after the reaction.

Tip: Upon separating the product after the magnesium combustion, a green substance is visible: magnesium nitride. The metal reacts with the nitrogen in the air by the very strong exothermic reaction with oxygen. If one drops water on the magnesium nitride, it changes to magnesium hydroxide and produces strong smelling ammonia gas.

E3.10 Conservation of Mass by Burning Charcoal

Problem: Students know from grill parties that charcoal is used for delivering the energy to heat meat or sausages on the grill. After looking to the burnt coal only little portions of ashes remain, and students like to think of a "destruction" of coal because they cannot observe the reaction with colorless oxygen and the colorless and gaseous product carbon dioxide. So it seems to be important to weigh first coal and oxygen in a flask, to burn the coal in the closed flask, and to weigh the flask afterwards. Observing the same mass as before the student may suppose that a colorless gas may be produced. The "lime-water test" can prove carbon dioxide.

Material: Analytical balance, big round flask with stopcock, rubber balloon; charcoal pieces, oxygen (lecture bottle), calcium hydroxide solution (limewater).

Procedure: Fill a flask with pure oxygen replacing the air, add 3–5 small pieces of charcoal, and close the flask with a balloon. Weigh the flask with balloon. Heat the flask so that the coal pieces will be ignited and shake the flask during combustion. After cooling down in the air, weigh the flask again and compare. Fill a few ml of limewater in the flask and shake.

Observation: A special mass of the flask is observed first. The pieces of coal are burning with very bright light, they disappear completely. After burning, no new substance is to be seen, but the mass of the flask is the same as before. The limewater shows the well-known white precipitation: the flask is filled with carbon dioxide.

E3.11 Burning Candles on a Balance

Problem: Students may understand the increase of mass in an open system during combustion of metals and the formation of solid metal oxides. They are not convinced that the same is true for alcohol, paper or candles from their everyday experiences. In order to show it practically, one can place a lit tea-light (candle) on the balance; in addition, carbon dioxide and steam can be captured with sodium hydroxide or soda lime (a mixture of sodium hydroxide and calcium oxide). In the apparatus (see picture), the gases are absorbed after

3.7 Experiments on Substances and Their Properties

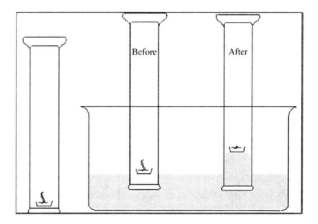

combustion; and the total mass increases because of the reacting oxygen. One should also discuss that the conservation of mass can be observed if candle and oxygen are weighed in a closed system and compared with the mass after combustion (see E3.10).

Material: Glass bowl, glass cylinder, glass plate, beaker, digital balance, glass tube with metal net, and funnel with glass tube (picture); soda lime or sodium hydroxide, tea-light, limewater.

Procedure: (a) Light a candle and place a beaker over it. Repeat this experiment using a cylinder. Extinguish the candle and add limewater to the cylinder; cover it with a glass plate and shake. **(b)** Put a cylinder over a burning

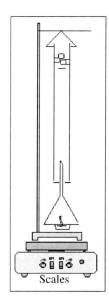

tea-light (without aluminum container), which is swimming in the water of the glass bowl. (c) Fill the upper side of the glass tube with sodium hydroxide so that formed gases can easily flow through it (see picture). Hang the glass tube on the balance, light the tea-light and observe the readings. If smoke develops (the absorption agent is too closely packed!), then repeat the experiment using less sodium hydroxide.

Observation: (a) A residue quickly forms inside the beaker: condensation of steam to little water drops. The candle flame goes out after a short while. The flame also extinguishes in the cylinder, the limewater turns milky upon shaking: colorless carbon dioxide. (b) The flame extinguishes, at the same time the water level rises inside of the cylinder. (c) The reading on the balance with burning tea-light rises showing the increase of mass up to 200 mg. The entire mass of both produced substances water steam and carbon dioxide is larger than the mass of the reacting tea-light portion before.

Tip: The raised water level in the cylinder after the flame is extinguished (see (b) in E3.11) cannot be explained through the "use of oxygen" because a correlative volume of steam and carbon dioxide is formed. The condensation of steam and the reduced gas volume through the cooling process of the gas after the flame is extinguished are responsible for the increase of the water level inside the cylinder.

E3.12 Reaction of Carbon Dioxide with Magnesium

Problem: At first, students accept that one can regain metals from solid metal oxides (see E3.3 and E3.4). However, they cannot imagine that it is possible to regain the carbon as a black solid matter from the colorless carbon dioxide gas. In order to convince them, one could dip a piece of burning magnesium in carbon dioxide gas. Students may suppose that burning magnesium will be extinguished by carbon dioxide like burning wood or candles – and will be surprised to see magnesium burning in pure carbon dioxide and have to solve this cognitive conflict.

Material: Cylinder, glass plate, crucible tongs; magnesium ribbon, carbon dioxide gas, sand.

Procedure: Place a 1-cm layer of sand in the cylinder as a protector for the bottom of the cylinder. Fill the cylinder with carbon dioxide by replacing the air and cover it with a glass plate. Hold a 10-cm piece of magnesium ribbon using crucible tongs, carefully light it and place it deep within the cylinder, taking care not to touch the cylinder.

Observation: The flame does not extinguish but continues burning with a crackling noise. On the inside of the cylinder, one can see black dots. When retrieving with one's finger, it is obvious it is soot that can color white paper black: carbon.

3.7 Experiments on Substances and Their Properties

E3.13 Density of Air and Carbon Dioxide

Problem: Students are surely aware of the atmosphere of air which surrounds us and our earth. To a much lesser degree, they identify air as a space-filling substance or as a mixture of substances with a characteristic and measurable density. This density should be established and compared with other gases. The density of air and its connection to sea level measurements can also be discussed in connection with the decrease of air pressure by climbing high mountains.

Material: Analytical balance, round flask with stopcock, syringe, vacuum source (water aspirator or vacuum pump), test tube; carbon dioxide (lecture bottle).

Procedure: Fill the syringe with exactly 100 ml of air. Evacuate the flask using vacuum pump and weigh. Attach the syringe to the flask and transfer 100 ml air from the syringe into the flask. Weigh again and calculate the air density. Repeat the experiment using carbon dioxide gas.

Observation: 100 ml air weighs 0.13 g. 100 ml carbon dioxide weighs 0.2 g. The densities can be calculated to 1.3 g/l (literature value 1.29 g/l) and to 2.0 g/l (literature value 1.97 g/l).

Tip: This experiment can also be carried out using a plastic bottle with stopper and stopcock: one weighs the bottle filled with air and attached stop cock, presses 100 ml of gas into the plastic bottle and weighs again.

E3.14 Properties of Hydrogen and Other Colorless Gases

Problem: Students tend to identify colorless gases as air, without critical reflection. For this reason, it is useful to demonstrate several colorless gases and look at their corresponding reactions, which would clearly show the characteristics of the various gases.

It is especially necessary to place emphasis on carbon dioxide gas, which can cause death. It is produced through the fermentation process by gaining ethanol from sugar solution and may fill the wine cellar by it's high density compared to air (see E3.13). Persons entering the cellar cannot breathe: the air is displaced by carbon dioxide. Therefore, such a cellar has to be tested with a burning candle before one enters it.

Because hydrogen properties and the imminent danger of explosions of hydrogen–oxygen mixtures, experiments on hydrogen should be demonstrated with care.

Material: Five cylinders, glass plates, wood splint, balloons, combustion spoon, test tube, beakers, empty can with concentric hole of approx. 1 mm diameter; lecture bottles of oxygen, nitrogen, carbon dioxide, hydrogen and butane gas, candle (tea-light), limewater.

Procedure: Fill the cylinders with the gases replacing the air, cover and label. (a) First place a burning wood splint in all cylinders and then a glowing one.

In order to differentiate between nitrogen and carbon dioxide conduct the following tests: **(b)** pour the contents of both cylinders into beakers, which contain a burning candle, **(c)** mix them with limewater and shake.

Observation: **(a)** The wood splint burns in air slightly, but very brightly in oxygen. Both the burning and the glowing wood splint are extinguished in nitrogen and carbon dioxide. Hydrogen starts burning with a bang and a colorless flame; one briefly sees a layer of water droplets. Butane lights up and burns with a yellow flame. **(b)** Carbon dioxide can extinguish a burning candle in a beaker due to its high density; nitrogen will not. **(c)** A white milky substance is formed from the colorless limewater in the presence of carbon dioxide, not of nitrogen.

Procedure of Further Hydrogen Experiments: **(a)** Fill a balloon with hydrogen, tie and release. **(b)** Bring a burning candle attached to a combustion spoon close to the balloon until a reaction sets in (Caution: balloon bursts!). **(c)** Light the streaming hydrogen from the lecture bottle and have a small flame. Place a dry beaker over it. **(d)** Fill an upside-down cylinder with hydrogen, introduce a burning candle attached to a combustion spoon. Slowly remove the candle and then re-introduce again, repeat this procedure several times. **(e)** Fill hydrogen in an upside-down cylinder, fill another same-sized cylinder with air and place it underneath, and mix both gases. Finally separate the cylinders from each other using glass plates and examine with burning wood splint (Caution: whistling bang!). **(f)** Place a can with a hole in an inverted position. Fill the can with hydrogen by replacing the air. Light the released hydrogen gas at the hole and wait (Caution: loud bang!).

Observation: **(a)** The balloon rises immediately. **(b)** The balloon bursts; the gas burns like a bright fireball. **(c)** The pure hydrogen burns quietly, the beaker is covered inside with moisture. **(d)** The candle is extinguished in the cylinder, lights up again when taken out. **(e)** The mixture of hydrogen and air burns very quickly with a loud bang. **(f)** At first, the hydrogen burns very lightly (one could place some paper over the hole as a control: it burns), after about 20 s one can hear a light burning noise and shortly afterwards a strong bang; the gas mixture burns very quickly: hot steam is formed and mixed with air.

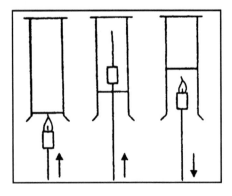

E3.15 Composition of Air

Problem: Students use the terms "good air" and "used air" in everyday language – they do not however think of the 20% oxygen level of fresh air and of changes in the percentage by reactions of oxygen. For this reason, it is important to carry out experiments demonstrating the composition of air, i.e. by reactions of metals or phosphorus with air. The question of using metals and phosphorus is discussed and explained in terms of the formation of *solid oxides* which "removes" oxygen from the air. If carbon or sulfur would be used, then *gaseous oxides* would form corresponding to the oxygen volume.

Material: Two 100 ml syringes, combustion tube, glass tub, small cylinder, glass plates, wood splint, glass cylinder with combustion spoon and stopper, ruler; steel wool, phosphorus.

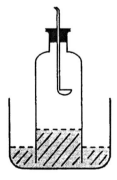

Procedure: (a) Assemble a combustion apparatus setup (see picture). Force a portion of 100 ml of air over heated steel wool several times and establish the volume of the remaining cooled-down gas. Collect the remaining gas pneumatically in small cylinder and test with burning wood splint. (b) Place an open glass cylinder in the water of a glass tub (see picture). Put a small portion of phosphorus in a combustion spoon, which has been pierced through a matching stopper. Light the phosphorus, take the spoon from the cylinder and close the cylinder with the stopper. After the flame is extinguished observe the increase in water level of the cylinder, estimate the volume of the left over gas using a ruler.

Observation: (a) The steel wool glows and produces a black substance: iron oxide. The gas volume decreases to about 80 ml and this remaining gas suffocates a burning wood splint: nitrogen. (b) Phosphorus continues burning for a while forming white smoke: phosphorus oxide. The flame extinguishes, the water level in the cylinder rises, the volume of the remaining gas is approximately 80% of the volume before: nitrogen. The white smoke dissolves in the water.

Tip: The solution of the phosphorus oxide in water can be checked using indicator solution or indicator paper. It is acidic due to the formation of phosphoric acid.

The percentage of 20% by the volume of oxygen in air can be discussed relating to the breathing process of people and animals. The carbohydrates and fat are "burnt" in every's body by the reaction with oxygen, the air which leaves the lungs shows the percentage of about 15% by the volume of oxygen, but nearly 5% of carbon dioxide. The plants are reacting with carbon dioxide and produce oxygen again: the percentage of 20% will be reproduced!

References

1. Engida, T., Sileshi, Y.: Concept Cartoons as a Strategy in Learning, Teaching and Assessment Chemistry. Addis Ababa, Ethiopia, 2004
2. Puettschneider, M., Lueck, G.: Die Rolle des Animismus bei der Vermittlung chemischer Sachverhalte. CHEMKON 11 (2004), 167
3. Schloepke, W.-I.: Alchimistisches Denken und Schuelervorstellungen über Stoffe und Reaktionen. Chim.did. 17 (1991), 5
4. Barker, V.: Beyond Appearances: Students' Misconceptions about Basic Chemical Ideas. London 2000 (Royal Society of Chemistry)
5. Pfundt, H.: Urspruengliche Vorstellungen der Schueler für chemische Vorgaenge. MNU 28 (1975), 157
6. Barke, H.-D.: Strukturorientierter Chemieunterricht und Teilchenverknuepfungsregeln. Chem.Sch. 42 (1995), 49
7. Driver, R.: Children's Ideas In Science. Philadelphia 1985
8. Muench, R., u. a.: Luft und Gewicht. NiU-P/C 30 (1982), 429
9. Weerda, J.: Zur Entwicklung des Gasbegriffs beim Kinde. NiU-P/C 29 (1981), 90
10. Voss, D.: Der Gasbegriff in den Vorstellungen der Schueler und Schuelerinnen. Staatsexamensarbeit. Muenster 1998

Further Reading

Abraham, M.R., Williamson, V.M., Westbrook, S.L.: A cross-age of the understanding of five chemistry concepts. Journal of Research in Science Teaching 31 (1994), 147

Anderson, B.: Pupils' conceptions of matter and its transformation (age 12–16). Studies in Science Education 18 (1990), 53

Bar, V., Galili, I.: Stages of children's views about evaporation. International Journal of Science Education 16 (1994), 157

Bar, V., Travis, A.S.: Children's views concerning phase changes. Journal of Research in Science Teaching 28 (1991), 363

Beveridge, M.: The development of young children's understanding of the process of evaporation. British Journal of Educational Psychology 55 (1985), 84

Bou Jaoude, S.B.: A study of the nature of students' understandings about the concept of burning. Journal of Research in Science Teaching 28 (1991), 689

Butts, B., Smith, R.: HSC Chemistry students' understanding of the structure and properties of molecular and ionic compounds. Research in Science Education 17 (1987), 192

de Vos, W., Verdonk, A.H.: The particle nature of matter in science education and in science. Journal of Research in Science Teaching, 33 (1996), 657

Hesse, J., Anderson, C.W.: Students' conceptions of chemical change. Journal of Research in Science Teaching 29 (1991), 227

Johnson, P.: Children's understanding of substances. Part 1: Recognizing chemical change. International Journal of Science Education 22 (2000), 719

Johnson, P.: Developing students' understanding of chemical change: What should we be teaching? CERAPIE 1 (2000), 77

Johnson, P.M.: What is a substance? Education in Chemistry 33 (1996), 42

Krnel, D., Watson, R.: Survey of research related to the development of the concept of matter. International Journal of Science Education 20 (1998), 257

Meheut, M., Saltiel, E., Tiberghein, A.: Pupils (11–12 year olds) conception of combustion. European Journal of Science Education 7 (1985), 83

Russell, T., Harlen, W., Watt, D.: Children's ideas about evaporation. International Journal of Science Education 11 (1989), 566

Schollum, B., Happs, J.C.: Learners view about burning. Australian Science Teachers' Journal 27 (1981), 107

Séré, M.G.: Children's conceptions of the gaseous state, prior to teaching. European Journal of Science Education 8 (1986), 413

Stavridou, H., Solomondou, C.: Physical phenomena – chemical phenomena: do pupils make the distinction? International Journal of Science Education 11 (1989), 83

Stavy, R.: Children's conception of gas. International Journal of Science Education 10 (1988), 553

Stavy, R., Stachel, D.: Children's ideas about 'solid' and 'liquid'. European Journal of Science Education 7 (1985), 407

Stavy, R.: Children's conception of changes in the state of matter: from liquid (or solid) to gas. Journal of Research in Science Teaching 27 (1990), 247

Stavy, R.: Childrens ideas about matter. School Science and Mathematics 91 (1991), 240

Stavy, R.: Pupils' problems in understanding conservation of matter. International Journal of Science Education 12 (1990), 501

Fig. 4.1 Concept cartoon concerning the structure of water [1]

Chapter 4
Particle Concept of Matter

The particle concept presents fundamental advantages, but also certain difficulties to students. The common particle concept of matter associates for each pure matter a special kind of particle, mostly spheres as models: e.g. one sphere for one water particle, another kind of sphere for one sugar particle, the mixture of both kinds of spheres the dissolution of sugar in water. On the other hand, one can use a generic particle term for atoms, ions and molecules; one can only ascertain it from the context which particle term is discussed.

In addition, students gain their first insight into the general use of scientific models through the common particle concept. Unfortunately these scientific models of matter have almost nothing to do with well-known models of cars or airplanes, with dolls or soft toys. There are originals for these models which can be built in miniature according to the subjective interest of the model builder: modeling a car one builder puts emphasis perhaps on the wheel-turning ability, the other one on glass windows of the model. For creating scientific models according to the composition of matter there are no atoms, ions or molecules to be seen, neither the chemical structure of matter nor chemical bonding – scientists have created these models according to the interpretation and reflection of many properties and chemical reactions of many substances.

Main Attributes of General Model Concepts. After an empirical analysis of the general model concept, Stachiowiak [2] differentiates between three basic traits. The *Image Trait* deals with "models that are always images of something, therefore representatives of certain natural or artificial origins". The *Shortening Trait* is about "models that do not represent all aspects of the original but only those which are relevant to the individual creator or user". The *Subjective Trait* means, "models fulfill their representational and substitution usage only for certain subjects being limited by particular theoretical or real operations".

Traits of the Scientific Model Concepts. Steinbuch [3] presents a scheme for indicating the epistemological process in science (see Fig. 4.2): "Any complex *reality* issue as the original, is recreated through a particular perception as an *abstract model*, a thought model by using only the essentials which are relevant. For this purpose certain information or generally-acknowledged rules of logic or physics are added. So we have for our perception a model for future thought

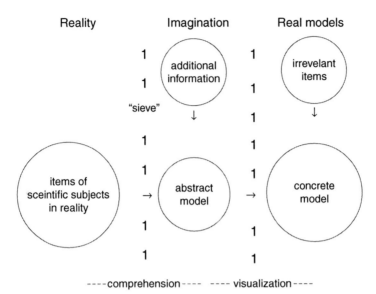

Fig. 4.2 Scheme "Thinking in model terms" by Steinbuch [3]

processes. This abstract thought model can be used for projection back to reality by building *concrete models*. These models contain unavoidable irrelevant attachments which the thought model do not contain" [3].

This "thinking in models" can be relayed for example as in the perception of Max von Laue, who in 1912 confirmed the structural theory of 3-dimensional crystal lattices by using a beam of X-rays [4]. The interference pattern of a sodium chloride crystal, which through interference and diffraction of the X-ray-beam is formed, is the original, and therefore the essential part, passing through the "sieve" (see Figs. 4.2 and 4.3).

Diffraction of light in two-dimensional lattices and their calculations were known in Laue's time: an *additional information*. They have been the basis for Laue's calculation of three-dimensional diffraction lattices from X-ray experiments; as a result he formulated a model of a spatial symmetrical structure of ions in a salt crystal: *abstract mental model*. Laue proposed the use of realistic models in order to better visualize the concepts – but needed irrelevant items like balls, sticks and glue, in order to construct closest packings of spheres or spatial lattice models: *concrete models*.

This scientific mode of moving from **"left to right"** in Steinbach's scheme is possible for an expert but not for a novice who is just beginning to think in terms of models. Students are much more likely to be led from **"right to left"** after having been introduced to the phenomenon or to the original. It is necessary initially to have concrete models in order to be able to develop further mental

4 Particle Concept of Matter

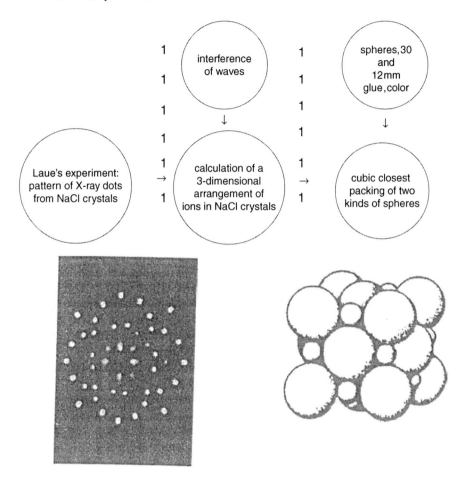

Fig. 4.3 Scheme "Thinking in model terms", example of Laue's path of perception [4]

models and to successfully come to appropriate models relating to the structure of matter. For instance, one could use close-packing models in order to demonstrate and discuss the structure of a common salt crystal and the sizes of the involved ions; i.e. the material and color of the spheres, glue and gaps in the packing should be recognized as irrelevant details. Afterwards one could demonstrate the crystal lattice model which only shows the positions of the involved ions; again the material or color of the balls and the sticks are irrelevant. Through discussion and comparison of these varied models one can give the young people an idea of the arrangement and sizes of ions in sodium chloride crystals (see also [4]).

4.1 Smallest Particles of Matter and Mental Models

"Water has no particles – a drop can be smeared at will; magnesium-particles are destroyed in combustion and ashes remain; sugar particles disappear when sugar dissolves in water – only the water tastes sweet".

These statements made by pupils after several hours of lessons drive every teacher to despair and make him or her wonder what went wrong in his or her teaching about small particles. The introduction of the particle concept is and remains difficult and cannot be mastered within a few school hours. Beginners start slowly and with many questions, arise at their first model concept of particles and their arrangement.

Even if the particle model of matter is introduced and well developed in chemistry lessons, one should never assume that children accept this concept and apply it to any type of matter. As in the times of Aristotle and Democritus, questions often arise regarding continuous or discrete (discontinuous) nature of matter. Learners have a problem of consistency of thinking, that is, in one context they think that particles exist and in another context there are none.

Teaching and Learning Suggestions. The ancient scientist Kepler accomplished a great theoretical achievement at the beginning of the 17th century. He observed six-sided snowflakes, developed the existence and arrangement of water particles and illustrated them with model drawings (see Fig. 1.2). The historical fact points out that by using a crystal, it is possible to theoretically follow the path to the smallest particles of matter.

Therefore pupils should prepare their own homegrown alum crystals before they are introduced to the particle model. They should initially observe that each crystal obtained from the saturated solution, whether small or large, has the same octahedral form (see E4.1). One can then discuss how to explain that resulting crystals always have straight edges and equal angles, without anyone having "used a ruler to measure or a file to produce the crystal". Perhaps the pupils will come up with the conclusion that each crystal contains small alum particles, which are arranged by nature with a specific construction plan. Should the kids not come up with this idea themselves, the teacher needs to bring it into the discussion.

One possibility would be to introduce a concrete model (see E4.2) and to compare it to the original (see Fig. 4.4). This comparison leads clearly to the association, that there are tiny particles in an alum crystal which are arranged in a special type of chemical structure like spheres in the shown closest sphere packing: original crystal and the concrete model exhibit the same octahedral form, the same smooth edges, smooth surfaces and uniform angles, one sphere correlates to one alum particle (see Fig. 4.4).

In order to develop a fact-based concept, it is also necessary to discuss the model's irrelevant components (see Fig. 4.2). In the depicted model the spheres are made of white cellulose; they could also be made of styrofoam, wood or plastic, neither material or different colors have a relevance to the model. Even the glue which binds the spheres has nothing to do with the crystal, there must be

4.1 Smallest Particles of Matter and Mental Models

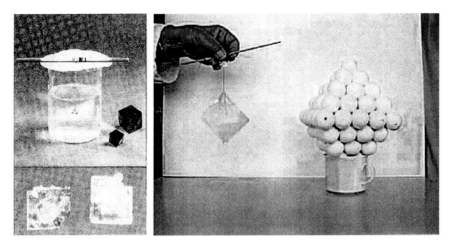

Fig. 4.4 Alum crystals in saturated solution, alum crystal and simplified structural model

force between the alum particles in the crystal in order to hold them in place. One can simulate such electrostatic forces by rubbing a plastic rod or an overhead foil with wool and touching pieces of paper with it (see E4.3). They are attracted by the electrostatic charge of the plastic rod or the overhead foil and stick to it.

In order to clearly show the irrelevant components, it is advantageous to have a second model available which has a different material for the spheres, a different color or a different glue. Perhaps it is even possible to create a virtual revolving computer image of the sphere packings enabling one to view them from all sides [4]. Virtual sphere packings are more fitting as a concrete model because one doesn't use irrelevant materials such as spheres, glue or sticks.

The picture of the closest sphere packing for the alum crystal (see Fig. 4.4) correlates to a rough educational reduction. The expert knows that the substance potassium aluminum sulfate dodecahydrate is composed of hydrated potassium ions, whose octahedral and tetrahedral interstices are filled with aluminum ions and with sulfate ions. This model concept may be introduced later and can be used in future lessons. Nevertheless, it is considered reasonable, at first, to introduce the "alum particles" and to choose reduced concrete models.

If one expects a pretty accurate and scientific model instead of the didactic model of the alum crystal then one could show an experiment demonstrating the crystallization of silver from silver nitrate solution by dipping in a copper wire (see E4.4). Then, one could also accurately compare and contrast the natural arrangement of silver particles in a silver crystal [4] using the model of the cubic closest sphere packing (see E4.5).

The first encounter with the particle model can also act as the basis of mental models of solutions, one is then automatically dealing with *two* particle types. As an example for this type of solution one could use substances that are composed of molecules: sugar, water, ethanol, etc. Salts are not suitable because

Fig. 4.5 Dissolving iodine in ethanol to form a brown solution [5]

they are composed of at least two ion types and cannot be easily described using the simple particle model.

A well-known example for young people would be the dilution of sugar in water. It is more spectacular however, to take iodine and to dilute it in ethanol, one gets a brown-colored solution. If one places an iodine crystal in a glass capillary and dips it in ethanol (see E4.6), a fine stream of the solution will escape from the capillary and will sink down due to its density. Only after quite a while does the iodine spread evenly through the entire volume (see Fig. 4.5).

Two types of spheres are used for a model experiment i.e. cellulose spheres of two different diameters (see E4.7). Little spheres are spread in a bowl which is positioned on the overhead projector. Several big spheres are at first arranged and packed in a regular and orderly fashion in the bowl, and then shaken together with the little spheres. All spheres are in constant movement and mixed to form a model for the iodine solution. The characteristics of this model are discussed. A big sphere represents an iodine particle, a little sphere an ethanol particle, the moving spheres are showing the moving particles in the solution. The irrelevant items are pointed out: neither the color nor the form are realistic, the movement of the glass bowl with the hand is irrelevant because the particles in the solution are moving constantly and independently by themselves. The model experiment is repeated through model sketching; finally, both are compared (see Fig. 4.6).

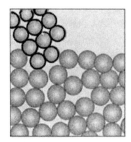 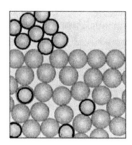 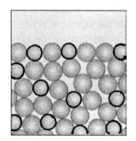

Fig. 4.6 Mental model visualized for the dissolution of iodine in ethanol [5]

Through such introductions to the particle model, young people may develop an awareness of the abstract mental model from concrete models: According to Steinbuch, students should be guided from "right to left" of the diagram (see Fig. 4.2). Should mental models for the structure of crystals or solutions exist, one could choose experiments and phenomena for the following lessons which are suitable for interpretation with the particle model: changes of states of matter, solutions, diffusion, extraction, distillation, etc.

Chemical reactions of metals to form alloys or of certain elements to form binary compounds cannot be used for the introduction of the smallest particles. The synthesis of water, hydrogen chloride or ammonia from the elements only make sense if the **Dalton atomic model** is introduced. For each element, a specific sphere is used as a model for this type of atom. The well-known molecular building sets are set up based on these principles.

The popular redox reactions for the formation of metal oxides, sulfides or halides from the elements even require a **nuclear atomic model** for the changes from atoms to ions. They are not at all suitable for introduction or use of the particle concept because the limits of the particle model would be way exceeded [4].

4.2 Preformed and Non-preformed Particles

"There are particles in ice crystals, when the ice melts, these particles disappear; sugar particles exist in the crystal sugar, but not in the sugar solution; gasoline particles are in liquid gasoline, when gasoline evaporates, they are destroyed".

Even though young people accept particles of matter in a discussion, difficulties always arise: the particle concept is not consistently used. Helga Pfundt [6] demonstrated the dissolution of a blue copper sulfate crystal in water and, after teaching them about the particle concept in chemistry lectures, gave them a questionnaire asking about their ideas (see Fig. 4.7).

Pfundt [6] differentiates existing answers not only in regard to a continuous concept or discontinuous concept, but also in regard to the possibility that particles are first created in the solution process or that during the crystallization process, existing particles join again forming continuous material. In this case, they are called **"non-preformed particles"**: they can appear and disappear. In other cases they exist all the time and are called **"pre-formed particles"**.

Pfundt lists two models for a salt crystal in the questionnaire (see Fig. 4.7): the upper model looks at the continuously built crystal; the lower model looks at a discontinuous crystal made up of particles. Three models are offered and compared with analogies for the solution process (from above to below): 1. without any particles ("like the way a drop of ink spreads in water"), 2. through the development of particles in solution ("like a piece of crystal sugar as it is ground to powder form"), 3. through the separation of existing particles ("like the way a lump of sand falls and separates to grains of sand").

The results show that grades 7–9 students mainly chose answers according to the continuous concept or thought that particles could be created by dissolving

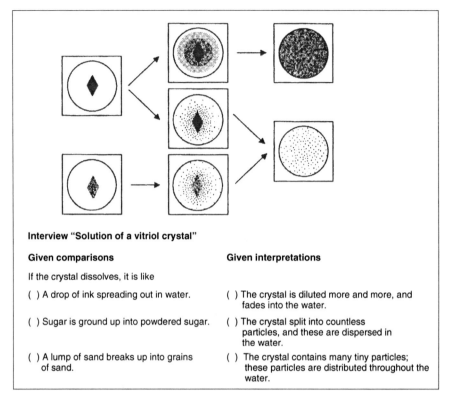

Fig. 4.7 Questionnaire based on the particle concept for the solution process [6]

crystal. Only a few of the students consistently chose metal models for preformed particles and consistently argued with the particle concept of matter.

Teaching and Learning Suggestions. It's a good idea to start chemistry lessons by demonstrating lots of interesting phenomena and surprising experiments and waiting a while before introducing the particle concept of matter. For example it is possible to introduce chemical reactions by the formation of oxides or sulfides from their elements – which would deliberately not be interpreted at the level of the particle concept. When the particle concept is later introduced in class, it should be consistently used for many subsequent subject areas and should be reinforced using different concrete 3-D models and 2-D model drawings.

Should the closest sphere packing model be used for a silver crystal (see Fig. 4.4), one could demonstrate the simulation of the melting of silver: the sphere packing – without using any glue – could be destroyed and single spheres shaken in a beaker. If the sphere packing is again layered, it becomes clear that even in the crystallization process of silver crystals from the molten matter,

4.2 Preformed and Non-preformed Particles

silver particles are retained: they do not "melt together", they are not "created" by the melting process. At this level, pupils can understand the popular model drawings of the states of matter – they only need to mentally add the movement of particles (see Fig. 4.8).

In the same way, the dissolution process of crystals in water can be consistently interpreted with adequate models. Should the cited sphere packing be used as a model for a sugar crystal, and another bunch of spheres as a model for a portion of water, it is possible to demonstrate the model of the solution by mixing and moving both sphere types (see Fig. 4.6). In the solution both water particles as well as sugar particles remain. By heating the sugar solution slowly, sugar crystals are again formed and water evaporates: water particles leave the solution and mix with air particles in the air. Sugar particles separate from the water particles and form a regular and symmetric pattern – both kinds of particles remain.

Further experiments and model drawings should follow e.g. the observation of the spreading of perfume in a room and the interpretation of the mixture of the continually moving perfume particles and air particles. Or the formation of snow and hoarfrost at winter temperatures below 0°C is interpreted via water particles which form regular pattern in snow flakes: they consistently have the six corners, never five or seven (see Sect. 1.6).

Students should see appropriate 3-D models and draw related 2-D model drawings in their notebook – so they will construct their own mental models. This way the particle model of matter can be utilized and internalized.

State of matter	Solid	Liquid	Gaseous
Arrangement of particles	Regular	Irregular	Irregular at all
Distance between particles	No distance	No distance	Large distances
Motion of particles	Vibrating at fixed places	Vibrating/changing their positions	Moving very quickly and colliding
Forces between particles	Very strong	Strong	No forces
Representation by the particle concept		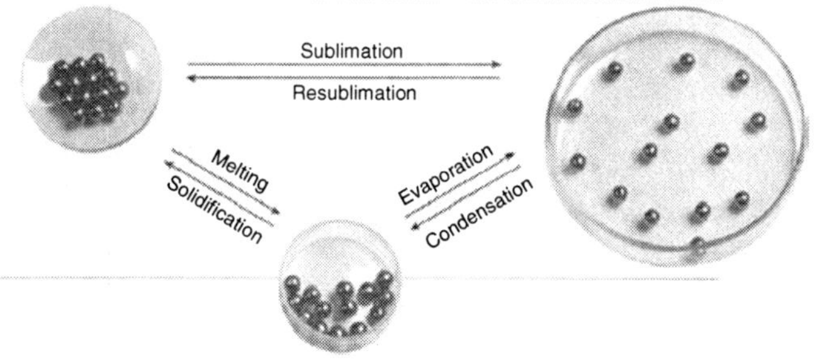 Sublimation / Resublimation / Melting / Solidification / Evaporation / Condensation	

Fig. 4.8 Concrete and mental models for the three states of matter [5]

4.3 Smallest Particles as Portions of Matter

"Sulfur particles are yellow; sugar particles are sweet; water is a fluid and consists of liquid particles; ice particles are solid; carbon particles burn in the barbecue, they smolder and turn to ash; the smallest copper particles are the smallest possible copper portions".

These and similar statements are to be expected from children after they first learn about model concepts but lack the language skills to adequately describe them. They mix up terms from the macroscopic area of matter like color, density, melting point or solubility, and the sub-microscopic area of the smallest particles like size or mass of particles (see Sect. 2.5).

Mixing these levels of terminology is difficult to avoid when the issue of smallest particles is introduced arising from the question of the division of a portion of matter: "is it possible to repeatedly and endlessly divide a piece of copper"? If the answer at the end of the discussion is that there are limits, then the result usually shows a smallest possible matter part. The ancient Greeks in Democritus' circle spoke of "atoms" probably meaning the smallest "indivisible" part of matter. If one critically researches the earlier literature, one will probably also see many of our chemists and physicists transferring the well-known characteristics of matter to atoms or molecules.

It is well-known that diamond and graphite are totally different substances with dramatically different properties (see Fig. 4.9), but are however composed of the same particle types, of carbon particles. One has to stop from transferring the material characteristics to the smallest particles. The carbon particle cannot be simultaneously "black" and "colorless"; it does not simultaneously have two "different densities"! It was only through X-ray structural analysis of the 20th century which finally proved, that both carbon modifications can be differentiated through distinct chemical structures. The different arrangements of the carbon particles in diamond and graphite are responsible for the macroscopic characteristics (see Fig. 4.9).

Naturally one cannot blame students if, after their first introduction to the particle model of matter, they associate certain colors with certain particles – especially if they have been using the molecule building sets in which green spheres are always used for the chlorine atom, the yellow spheres for sulfur

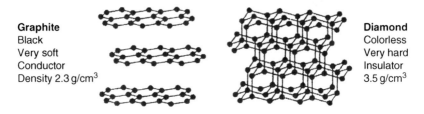

Fig. 4.9 Diamond and graphite: characteristic properties and chemical structures

atoms and the black spheres for carbon atoms. One should not be surprised about their ideas, these misconceptions are mostly school-made!

In order to qualify the material characteristics of "color" and in order to avoid transferring these characteristics of matter to particles the use of colored spheres should be avoided in the lesson. If one uses different colors and materials for the spherical models it is necessary to hold critical model discussions and to name the irrelevant items (see Fig. 4.2).

Teaching and Learning Suggestions. As in the teaching-learning suggestions of the previous sub-chapter (see Sect. 4.1), one can use the close packing or a crystal lattice model as an observational 3-D model to show the crystal structure. The same applies for crystals of diamond and graphite. Both crystal structures clearly show defined differences in the arrangements of the carbon particles (see Fig. 4.9).

One should bring across the fact that a tiny crystal which is barely visible through a microscope and which weighs a mere few milligrams contains an unbelievably large number of particles. If a diamond of a mass of 12 g contains the tremendous amount of 1 mol C atoms or 6×10^{23} C atoms, then the tiny crystal of 12 mg still contains 6×10^{20} C atoms. Structural models for diamond crystals, which are formed from 20 or 50 spheres in the form of a crystal lattice, differ quite a bit from reality. They merely portray a tiny part of the specific structure of billions upon billions of C atoms in a visible crystal.

If it is said to be possible, under very high temperatures and pressures, to create small diamonds of a greater density from the graphite crystals, then the following becomes more apparent (see Fig. 4.9): there are no specific diamond or graphite particles! Carbon particles form both modifications; graphite crystals in layered structure (that is where the lubricant characteristic comes from) and diamond crystals in tetrahedral structure and with other bonding relationships (that's where the greatest hardness of all substances comes from). The characteristics of a substance stem more from the spatial arrangement of their particles than from the type of particles involved, i.e. from the chemical structure!

An equally convincing example can be found in both phosphorus modifications. P atoms can, on the one hand, arrange themselves in groups of four (P_4 molecules) and can build a molecular lattice. For example, white phosphorus results with the lower melting temperature of 44°C and a density of 1.82 g/cm^3, it is self-combustible and very toxic. P atoms can, on the other hand, arrange themselves in a specific infinite layered lattice. For example, red phosphorus results with the higher melting temperature of 620°C and the density of 2.36 g/cm^3 – it is neither self-combustible nor toxic. The characteristics are determined by the arrangement of the P-atoms! If small portions of both phosphorus types react with oxygen, both produce white smoke. It is formed from the formation of phosphorus oxide crystals in air which when dissolved in water produce an acidic solution. These two forms of phosphorus produce the same oxide and the same acidic solution (see E4.8).

Apart from erroneous conclusions regarding the "colored particles", there are often misconceptions concerning the smallest particles, which could be "solid", "liquid" or "gaseous". With this in mind, single spheres can never convincingly to represent matter. It is possible to represent a portion of matter whether solid, liquid or gaseous by using an adequate number of spheres even for a small crystal or a tiny drop. In a model drawing 14 spheres have been used (see Fig. 4.8). This model does not only represent the regular symmetric formation of particles in a crystal but also formation in liquid and corresponding gases. The same particles are merely arranged in a different manner, they move independently and at varying speeds.

Let us now explore how and why volumes increase in the transition from liquid to gas or to steam. If one takes liquid ethanol and places a few drops in a balloon, closes it and dips it in the steam of boiling water, it will expand. It will shrink to its original size when cooled (see E4.9). Ethanol particles fill a much larger volume in ethanol steam than in the liquid. They do not get bigger, which students might at first think, but they move much faster. A correlative model should show both, i.e. volume increase and particle movement.

In the model experiment (see E4.10) the spheres are moved in the watch glass in such a way that they oscillate. A stronger movement in the Petri dish causes the spheres to stay together but constantly change sphere neighbors. A very strong movement in a large glass bowl causes the spheres to move at a high speed, the spheres fill the entire bowl with their movements. The drawing (see Fig. 4.8) shows the static model for the volume increase, but the model experiment (see E4.10) shows the dynamic model: the particles of gas move very fast, greater distances are created between the particles, they fill every available space and cause pressure to the wall of the vessel.

The movement of the glass bowl by the experimenter is irrelevant, the particles in a portion of matter move independently, the higher the temperature the more the movement. What is the substance between the particles of gas or steam? In the model one finds air between the spheres – an irrelevant item! In reality there is nothing between the particles of a gas.

4.4 Particles and the "Horror Vacui"

"Space between the particles cannot be empty, something has to be there; I cannot imagine that there is nothing; if there is no air, then it must be a vacuum, and I just cannot imagine that; something must exist, there is no place where absolutely nothing exists; the space cannot contain nothing; something has to be there!" [7].

In the model drawing showing the formation of matter, Pfundt (see Sect. 4.2) determined that children tended to choose the square or cube as a model for particles instead of the usual sphere. When asked for the reason they answered that the models "has to fit in such a way that they are connected to each other without leaving a gap" [6]. If one takes spheres they would have gaps which, in

4.4 Particles and the "Horror Vacui"

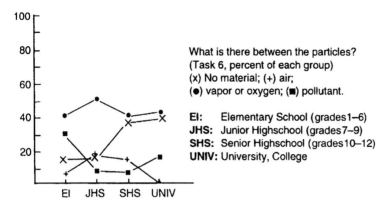

Fig. 4.10 Empirical results concerning the "horror vacui" by Novick and Nussbaum [8]

the opinion of the children, should not be there. The "horror vacui" in the imagination of the students led to the preference of cubes rather than spheres!

Novick and Nussbaum [8] also sent out questionnaires regarding the comprehension of particles and realized when talking of gases that the majority of students in the U.S.A. are of the opinion that air or other matter can be found between the gas particles (see Fig. 4.10). Based on this, further studies were carried out in Germany about the "horror vacui" and the space between the particles of gases [9].

Other empirical data were established using short experiments [7] – one experiment was carried out using butane gas (see E3.6). The appropriate model was requested in a questionnaire and questions were asked regarding the space between the butane particles (see Fig. 4.11). The results of the questionnaire in grades 9–11 show that almost all students draw the model drawing correctly. However, to the question regarding the space between particles only about 50% of the students mark the options of "nothing" or "empty". This means that the other half of the students assumes that spaces between the particles are filled with butane, with air or with other matter. These young people are victims of the "horror vacui"! This is obvious by the explanations which are given by the students (cited at the beginning of this section).

Apart from the introduction of the smallest particles, it is also necessary to discuss the inner space between the particles when discussing and teaching about the particle model: "The empty space is apparently missing in our particle model; this empty space however is the unbelievable aspect in the *discontinuum* concept, much more so than the denomination of matter" [10]. How can we change the lesson so that this unbelievable factor is convincingly accepted and understood by the students?

Teaching and Learning Suggestions. In order to understand the concept of spaces free of matter, one must first understand that air is a form of matter in the chemical sense, that air can be measured as having a specific density

Models for the evaporation of liquid butane (camping gas)

The cylinder of a gas pump or syringe is filled with colorless butane gas (see picture). The piston is pressed by hand and liquid butane appears. The piston can be released and liquid butane turns into gaseous butane.

Draw your mental model of the formation of butane particles:

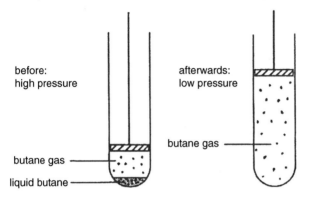

How do you imagine the space between the particles of the butane gas? I think......

( ) there is also butane gas between the particles,

( ) the space between the particles is empty,

( ) there is nothing at all between the particles,

( ) there is air between the particles,

( ) there is a special invisible substance between the particles.

Fig. 4.11 Part of a questionnaire (model drawing of particles by a student) [7]

(see E3.13). It is also possible to show combustion reactions in the air using closed apparatus, to differentiate between oxygen and nitrogen and to calculate the air content by measuring their volumes (see E3.15). The value 80vol-% nitrogen does not only show the amount of this gas in air, but also the fact that air is a mixture of different gases; that air and oxygen or nitrogen are substances, are in fact matter.

Now it is time to demonstrate air-free and matter-free spaces – the vacuum. If one demonstrates the mass of an air portion by weighing back an initially evacuated glass flask, then a central experiment has been carried out (see E3.13), i.e. one has experienced the practical evacuation of a flask. If one then conducts a discussion about the "vacuum-packed" food or "evacuated television tubes", it becomes evident that matter-free space can be created by pumping out the air. One could also reflect upon the possibility of a real vacuum by calculating the steam pressure of water through the use of water jet pumps or the partial pressure of oil in oil pumps: a remnant of steam of water or oil is always left after pumping out the air.

4.4 Particles and the "Horror Vacui"

In order to demonstrate the existence of vacuum one could show the students reduced air spaces and also spaces without any air. One could discuss the sinking of air pressure in a syringe by pulling out the piston and point out the resistance through the outer air pressure. Or one could replace the remaining air with a different type of gas in the flask: the air is thinned out even in this case. It is also possible to discuss the aspect of air-free spaces in two ways; either by creating vacuum or replacing the air by a different type of another gas.

Ultimately the difference in the comprehension of the terms "empty" and "nothing" should be made apparent. A beaker, which in the eyes of the students is pneumatically filled with air (see E4.11), is commonly known as an "empty glass" although it is filled with air. If one wanted it to contain "nothing", i.e. free of matter, one would have to close it off, attach a vacuum pump and pump off the air (see E4.12).

Expanding upon these experimental experiences with a vacuum, it makes sense to introduce the matter-free space between the particles of gases. It is possible in that context to enclose a liquid, which can easily be turned to vapor, i.e. ethanol in a balloon and to place it in a simmering water bath (see E4.9): a remarkable increase in volume is to be observed. If one takes the balloon out of the water, it shrinks back to the original size, i.e. the same portion of ethanol vapor condenses to a few drops of liquid ethanol of the same mass than before.

The children could have a follow-up discussion showing that the balloon is a closed-in container from which substance can neither enter nor leave. This means that the same ethanol amount exists before and after the condensation process. The same amount of ethanol particles has spread throughout the much greater volume of vapor; the particles have to move quite a bit, thereby creating larger distances between each other. One concludes that because the volume has increased by the factor of about 1000, the distance between the particles – the average mean free distance – amounts to 10 times as much as the average of the particle diameter. The interstices are formed by the fast movement of the particles; they are not filled by matter in any form. If one assumes that an electromagnetic field exists between the particles and wishes to discuss it in the sense of Einstein's equivalence of matter and energy, this can be individually decided upon [7].

One can further investigate the volume increase by using a model experiment for the particle movement of gases. The "instrument for kinetic gas theory", which sets small steel spheres in movement in a see-through cylinder, shows a larger volume of moving spheres by increasing the movement frequency of the spheres. This model however has to be discussed in two respects: 1. the gas particles move independently and need no "vibration motor" in order to move, 2. one finds no matter in the interstices of the original gas, however, one finds air in the model of moving steel spheres as an irrelevant item (see Fig. 4.2).

If one has no such "instrument for the kinetic gas theory", one could use a Petri dish almost filled with small spheres which can be manually moved through shaking the dish: model for particle movement in a liquid (see E4.10). In order to model the evaporation process, the spheres can be poured into a large glass bowl and strongly shaken: a model for particle movement in a gas.

The shaking and the air between the spheres are irrelevant ingredients as far as the discussion is concerned (see Fig. 4.2).

Despite all the methodically correct efforts, one must expect the horror vacui to show up with children as well as adults – the idea of "nothing" can be especially difficult to comprehend. An experienced academic advisor expressed this during a lesson by telling his students: "Just imagine the entire room here is filled with vacuum!" (Abitur paper of the year 1965 at the Gymnasium in Lehrte near Hannover, Germany).

4.5 Particles – Generic Term for Atoms, Ions and Molecules

"Iron particles form definite structures over magnets on glass plates", "the smallest particles of water are H atoms and O atoms", "hydrogen chloride particles contain chlorine and hydrogen".

The term for particles is much too flashy for most people. In everyday language, the name suggests a small portion of matter: chips of iron filings could also be called iron particles, crystals of powder sugar could also be known as sugar particles, small sulfur crystals also as sulfur particles: one therefore cannot reproach young students for using such synonyms in the particle terminology.

On the other hand, the term diversity increases quite a bit in regard to the particle after the subject of the atom has been introduced. Suddenly the H atoms and the O atoms become the "smallest particles of water", in reality the atoms are even smaller units than the relating molecules. Using this logic there is nothing more obvious to the students than that "hydrogen and oxygen escape when water is boiled".

Should the term "atom" be known and the term "chlorine atom" is shortened to the term "chlorine" which often happens amongst insiders using laboratory jargon, then it is possible that statements are made such as "chlorine and hydrogen are contained in particles of hydrogen chloride". The chemistry beginner and the critical student find this as ultimate confusion; they may think it over and consider how hydrogen (combustible) and chlorine (green and toxic) could be contained in hydrogen chloride (see also Sect. 4.3). Also the desired graphic descriptions introduce misconceptions to the term particle or atom in the sense of smallest units of matter.

Weninger [11] made the suggestion of using the word "monad" for the smallest particle of matter or the "16-proton atom" as the name for the sulfur atom with the intention of avoiding misconceptions concerning differences in matter and particles: sulfur should be the name of the matter, "16-proton atoms" the name of smallest particles of sulfur. However, these suggestions were much cited but never imposed. Buck [12] also opposes concrete models or graphical descriptions of particles or atoms and distinctively asks: "how can one teach the difference between matter and atoms?" [12].

Teaching and Learning Suggestions. The scientific particle term has to remain associated to the sub microscopic level, it should not be used for small portions of material. Of course, one can speak of iron filings, small sulfur crystals, tiny water drops or of little gas bubbles at this level; it should be no problem to avoid the particle term when talking about matter. Students will thereby realize that the smallest particle is reserved for the invisible area of mental models of matter, i.e. water particles, sugar particles, and ethanol particles (compare Sect. 4.1).

Besides the use of the particle term at the level of the particle model, it is also usual to use the particle as a generic term for atoms, ions and molecules on the level of Dalton's atomic model. One asks for instance, which particle types can be found in a salt crystal and expects the answer, that there are sodium ions and chloride ions. Should one ask about particle types in a sugar solution, the answer should be: "sugar molecules and water molecules".

According to the Dalton atom terms, it is ridiculous to ask what the smallest particles of water are and to accept H-atoms or O-atoms as an answer. The particle term should always be associated with units which independently move in a substance: Ne atoms in neon gas, Na^+ ions and Cl^- ions in the molten rock salt, H_2O molecules in water. H atoms and O atoms do not move independently from each other in water but are contained within H_2O molecules. In addition, the question of the smallest particles of water is a semantic contradiction – only the question regarding the number of atoms in one H_2O molecule can be answered correctly: "there are two H atoms connected to one O atom".

For students, it means being able to distinguish from the context, whether particles are meant in the sense of the particle model, or particles as a generic term for atoms, ions and molecules. If in addition to the protons, neutrons, and electrons the "elementary particles" come into play, then the term should be easily distinguishable in the right context.

4.6 Formation of Particles and Spatial Ability

"One sphere is touched by 6, ... no by 8 other spheres; ... I count 12 spheres, no I think there are 14 spheres...; there are quarter-spheres in the corners of the unit cell, no – I see them as eight-part spheres...or ...?".

If one lets the students build a close-packing model for a silver crystal (see E4.5), the stacking of the spheres is at first merely a mechanical problem. If, however, they have to find the coordination number, i.e. 12 for the number of spheres, which touch one central sphere inside of the packing, it becomes a problem of spatial ability [13]. One student might be able to visualize those six spheres surrounding the central sphere on one layer, three spheres above and three below (see E4.5). Another student might have to count out each individual sphere which visibly touches the central one.

In these cases there are no misconceptions, as previously discussed, but there are different developmental stages in spatial ability. In order to improve this

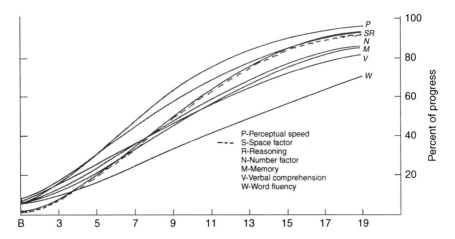

Fig. 4.12 Development of primary factors of intelligence [15]

ability through training, structural models of matter are ideal. Spatial models of crystal structures like sphere-packings, crystal lattices or unit cells are suitable, and improve not only the understanding of chemistry but especially spatial ability!

Teaching and Learning Suggestions. It has been known for quite some time that spatial ability is a primary factor of intelligence [14] and that this ability is developed in childhood [15] (see Fig. 4.12). In addition, many studies have shown that girls are particularly disadvantaged in this area (see Fig. 4.13). In Ethiopia it was found that specially boys from Private schools are better than the girls [16], in contrast boys and girls in Government Schools show the same ability in grades 7–10, but in grades 11 and 12, boys gain higher results [16]. The hypothesis remains that in Ethiopia wealthy parents of children in Private schools buy their children technical toys, which particularly help boys in their training for spatial ability. The poor parents of children in Government Schools cannot afford such toys; neither boys nor girls get any training, so spatial ability remains equal, but weak in comparison to children in the Private schools [16].

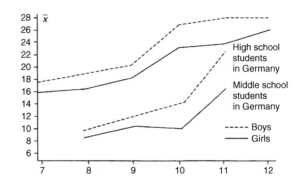

Fig. 4.13 Spatial ability in relation to class level and gender [16]

4.6 Formation of Particles and Spatial Ability

Many researches have shown that boys show significantly better results in spatial ability than girls (see Fig. 4.13). Rost [17] discovers in his studies, "that there exist stable gender-specific competence profiles, which show a slight superiority in girls regarding verbal competence and a slight superiority in boys regarding numerical and graphic representational forms, especially with regard to mental models" [17].

In order to compensate for these advantages of the boys in spatial ability, an early training with three-dimensional structural models in chemistry lessons could be especially helpful for the girls. With this in mind, structural models or model drawings mentioned in this chapter should be constructed and discussed (see Fig. 4.2). The observation of many other additional chemical structures will be examined in Sect. 5.

One could introduce a test which diagnoses the differences and successes regarding spatial ability for recognizing chemical structures [13]. Two assignments from such a test are demonstrated (see Fig. 4.14): students count the number of bricks and they should even consider the invisible bricks, they should also visualize taking bricks from a hollow space and mentally counting them. In the second assignment, the spheres of a packing unit should be counted, again invisible spheres are included: students have to picture these invisible spheres inside the unit and mentally determine the coordination number of 12 spheres touching one central sphere inside the packing.

There are further ways which demonstrate crystal structures and which challenge and train students to acquire spatial ability. For instance, there are red-green stereo pictures of metal and salt structures which, when looked at with red-green glasses, produce an amazingly spatial 3-D effect [18]. These pictures begin to develop in people's minds only after a few seconds and create wonder. There are also computer programs which show spatial effects through the rotation of chemical structures at various axes, an example is the multidimensional learning

1.3. How many bricks does this hollow wall contain?
 4, 6, 8, 10, 16, (20)

1.4. How many bricks do you need to fill the hollow wall?
 (4), 6, 8, 10, 16, 20

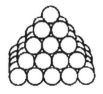

2.5. How many spheres does this sphere packing contain?
 30, 36, 50, (55), 56, 64

2.6. How many spheres are totally inside of the sphere packing?
 3, 4, (5), 6, 7, 8

2.7. How many spheres are touching one inner sphere of the packing?
 6, 8, 10, (12), 14, 18

Fig. 4.14 Examples of problems in spatial ability test, results in parentheses [13]

environment of "Metal and Salt Structures" which can be interactively used by students [19]. Spatial abilities can be vastly improved in various school subjects and professions through structurally oriented chemistry lessons and with different media and models on the topic of structure of matter.

Sopandi [20] empirically confirmed that there are even associations between the understanding of chemistry and spatial ability of students. Good spatial abilities of students in grades 9–12 in German Academic high schools around Muenster correlate with good test results in understanding chemistry [20]. Because there are special high significant differences in spatial ability between boys and girls moving from grade 9 to grade 10, Sopandi looked at the curricula and the guidelines of science education in these classes. He showed that in chemistry lectures, there are many content topics using structural models: metals and their structures, the structures of some salt crystals, the modifications of diamond and graphite, the spatial structures of many molecules in organic chemistry, etc. Because there are no significant differences of spatial ability in grades below 9 or above 10, one can deduce that the jump in spatial ability is due to using structural models in chemical education. Also Maccoby and Jacklin [21] and others confirm that spatial ability can be successfully taught.

It seems that spatial ability provides a big advantage for students from all over the world. The use of many structural models in chemical education means that students gain not only a good understanding in chemistry, but that they also develop good spatial ability for their profession – no matter what profession they later choose!

4.7 Diagnosis Test for Understanding the Particle Model of Matter

A teacher may ask his or her students about their understanding of the particle model of matter. If the teacher needs a written questionnaire, he or she could use a diagnosis test created by Kathrin Brockmann [22] at University of Muenster. She developed the test "Particles of Matter", utilizing some of the very well-known misconceptions held by most students. Finally, she evaluated this test with about 160 German students aging from 13–15 in the 7th grade [22].

Diagnosis Test "Particles of Matter"

Problem 1 *Mark one of the letters to complete the following sentence:*
If we could see the air particles in the flask (picture below), we would find out that space between the particles is filled with

(a) air.
(b) contaminants.
(c) oxygen.
(d) nothing, no matter at all.
(e) steam of water.

(f) dust.
(g) another invisible substance.

Problem 2 *A test tube is filled with butane gas and closed with a stopper. In the model drawing, the butane particles are marked by dots (picture below). Explain why the particles are distributed uniformly? Mark your answer:*

(a) Because there are other particles in the space between the particles.
(b) Because there is air in the space between the particles.
(c) Because there is butane in the space between the particles.
(d) Because there are repelling forces which keep them far apart.
(e) Because particles are moving randomly.

Problem 3 *A crystal of ice is built up by water particles. In the box, please draw a model of how water particles are arranged in a crystal of ice.*

Problem 4 *A balloon is attached to a flask that is filled with air (picture 1 below). The bottle is heated with a burner. The balloon inflates (picture 2 below).*

picture 1

picture 2

4.1 Why does the balloon inflate? What is the effect of heating on the particles of air in the bottle? Mark your answer:

(a) The particles of air are expanding and getting bigger.
(b) With heat, the number of particles is increasing.
(c) Distances between the particles increase and the particles are moving faster.
(d) The particles of air are moving out of the bottle into the balloon to get away from the heat.

4.2 Draw, in picture 2, your view of the particles of air after the bottle with air has been heated.

Problem 5 *Which figure shows a little crystal of sugar dissolving in water? Mark your answer:*

(a) The sugar crystal is dissolving more and more, and the sugar is mixing with water.

(b) The sugar crystal is dividing in a countless number of tiny sugar particles. These particles are mixing with water.

(c) The sugar crystal is built up by countless tiny sugar particles. In water, the sugar crystal is dividing in these particles. Sugar particles and water particles are mixing uniformly.

Write reasons of your choice:
_____
_____
_____
_____

4.7 Diagnosis Test for Understanding the Particle Model of Matter

Problem 6 *If you put an ice cube on a heated surface, you can observe that water can have the three states of matter: solid, liquid and gaseous. Which of the following model drawings of an ice cube on the heated surface matches best with your image?*

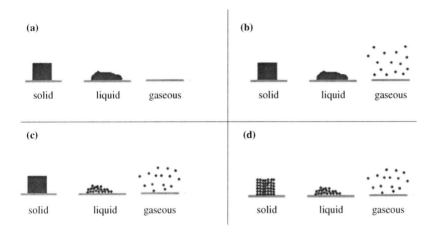

Write reasons of your choice:

_____

_____

Problem 7 *Mark all answers that are correct in your opinion:*

(a) In ice crystals the water particles are solid, in water they are liquid, in steam they are gaseous.
(b) In water, the water particles are blue – in steam, they are invisible.
(c) The water particles have no color at all.
(d) Water particles are moving. In steam they are moving fast, in water slow, in ice very slow.
(e) Water particles are moving in steam and in water, but in ice crystals they are not moving.

Findings

Brockmeyer [21] evaluated the test "Particles of Matter" with about 160 German students in age of 13–15 years in German grade 7 of the academic stream. She got the following results represented by bar charts, they are shown without comments.

Problem 1 asks the question "what is there between the particles". Answer D is the right one, about 58% of the students chose that answer.

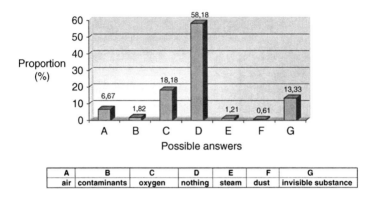

Problem 2 asks the question "why butane particles are distributed uniformly?". The right answer is E: "particles are moving randomly", about 59% of students chose that answer.

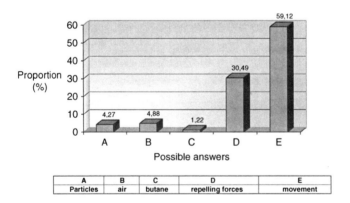

Problem 3 asks for a model drawing of the arrangement of water particles in an ice crystal. Nearly 75% of the students did it in a right way, 20% were wrong, only 5% did not use the particle model of matter.

4.7 Diagnosis Test for Understanding the Particle Model of Matter 91

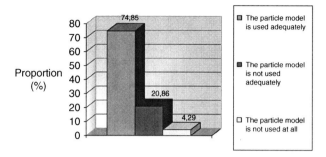

Problem 4 shows that a balloon gets bigger if a closed flask of air is heated. The question concerns the interpretation of the increasing volume by particles of air, the right answer is C: "distances between the particles increase and the particles are moving faster", about 53% chose it.

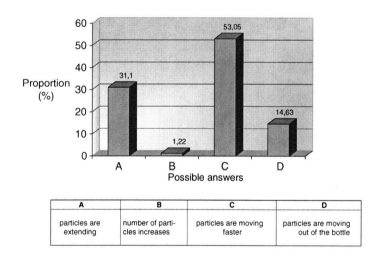

A	B	C	D
particles are extending	number of particles increases	particles are moving faster	particles are moving out of the bottle

Problem 5 proposes three kinds of mental models for the process of dissolving sugar in water: about 70% of the students got the right choice C.

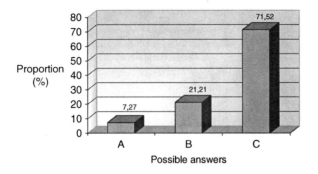

A	B	C
The sugar crystal is diluting more and more, mixing with water	The sugar crystal is dividing in sugar particles which are mixing with water	The sugar crystal is built up by sugar particles, they are dividing in water, are then mixing with water particles

Problem 6 offers four model drawings according to the states of matter solid, liquid and gaseous: about 90% marked the right answer D.

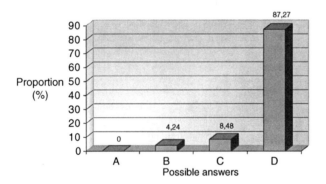

A	B	C	D
All states of matter are marked without any particles	Only the steam contains water particles	Water and steam contain water particles	Ice, water and steam contain water particles

Problem 7 shows five answers according to properties of particles of water: the right answers C and D are marked by about 90 and 70% of the students.

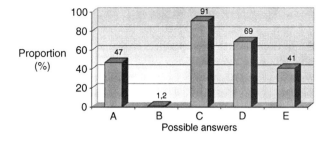

A	B	C	D	E
In solid ice particles are solid, in water liquid	In water the particles have the color blue	Particles of water have no color at all	Particles of water are moving in all three states	Particles of water are not moving in ice crystals

4.8 Experiments on Particle Model of Matter

E4.1 Growing of Alum Crystals

Problem: Students should be amazed at the clear lines, the smooth surfaces and the constant angles which occur in alum crystals which have been grown from the saturated solution – they may think that human hands have manipulated the crystals. In the discussion of these straight lines and smooth surfaces, students – with the help of their teacher – may realize that the alum crystal is composed of small particles that are arranged according to a particular construction plan. One should note that it takes several weeks before a crystal exhibits a few centimeters in its edge length – therefore the growth process must begin well before the lesson starts.

Material: Crystallizing bowl, 2 beakers, funnel, filter paper, glass rod, thread; alum salt $(KAl\,(SO_4)_2 \times 12\,H_2O)$.

Procedure: Dissolve approximately 40 g of alum in 200 ml of warm water, and cool the solution. Filter the saturated solution and take a little portion of it into the crystallizing bowl. Let it stand for two days until some octahedron-shaped crystals are formed. Attach the best crystal to a thread and hang it from the glass rod into the saturated solution filled into the beaker. The crystal is completely covered. After a few days remove new crystals from the thread, and filter the solution again. Occasionally add saturated solution of the same temperature.

Observation: The crystal grows to a fist size over weeks and months; it has the form of an octahedron. Its edges are perfectly straight, the surfaces are smooth, and the angles between the surfaces are constant.

E4.2 Close-Packing Model for the Alum Crystal

Problem: The explanation for the symmetric octahedral shape of the crystal lies in its chemical structure which can be described as a cubic packing of hydrated potassium ions whose octahedral and tetrahedral gaps are filled with hydrated aluminum ions and sulfate ions, respectively. If one assumes the simple particle model of "alum particles" and chooses the spherical model for one alum particle, then one can explain the simple structure as a cubic close packing of many similar spheres. It is possible to reconstruct such a sphere packing in an octahedral form and it can be recognized as a structural model when compared with the original crystal.

Material: 100 spheres (cellulose or styrofoam, white, 30 mm diameter), glue.

Procedure: Glue together a layer of 5×5 spheres in a square shape, in addition glue together further layers with 4×4 spheres, with 3×3 and with 2×2 spheres. Place these layers upon each other as demonstrated (see Fig. 4.3). Finish the arrangement placing one single sphere on top, and another sphere at the bottom.

Observation: Octahedral shapes of alum crystals and close-packing models are identical.

E4.3 Electrostatic Forces for a Bonding Model

Problem: The students are probably capable of visualizing and accepting the close-packing of glued-together spheres. However, they are bound to ask what attractive force keeps the particles together in the original crystal. If one rubs a new overhead projector foil (transparency) on a piece of paper and brings both together, both foil and paper attract each other. One can also rub a plastic rod on wool and use it to pick up little pieces of paper from the table. These attracting forces are known as electric forces which are responsible for bonding of particles in a crystal.

Material: Overhead projector transparency and paper, plastic rod and wool.

Procedure: Rub vigorously transparency and paper against each other, separate, and hold close to each other. Similarly rub two transparencies with paper in the same way and hold both transparencies close to each other. Repeat the experiments with plastic rod and wool: rub the plastic rod on wool and hold close to little snippets of paper on the table.

Observation: Transparency and paper attract each other, transparency and transparency repel each other, plastic strongly attract the paper snippets.

E4.4 Silver Crystals from a Silver Salt Solution

Problem: The close-packing sphere model for the alum crystal is an educational model, it is not correctly representing the facts. If a suitable model for an original crystal has to be constructed, one should choose metals with a cubic face centered structure, such as noble metals, aluminum, nickel or lead. A cubic close-packing model can be constructed and used for crystals of these metals. Because there are few convincing metal crystals to be found in nature, silver crystals should be grown by a reaction from the silver nitrate solution and observed.

Material: Test tubes, glass rod; copper wire, silver nitrate solution.

Procedure: Wind up a coil of copper wire around a glass rod several times forming a small wire helix. Remove the glass rod and hang the helix into the test tube. Fill the test tube with silver nitrate solution so that it covers the helix. Wait.

Observation: Silver-colored crystal needles appear on the copper wire: silver crystals. After quite some time the colorless solution turns light blue: copper nitrate solution.

Disposal: Crystals and solution mixture are disposed of in the container for heavy metal salts.

E4.5 Cubic Close Sphere Packings as Models for a Silver Crystal

Problem: The Ag atoms in a silver crystal are arranged like spheres in the cubic closest packing. Students may pack spheres of the same diameter as densely as possible, thereby casually modeling the correct structure of the silver crystal. The structure is called the *cubic* close packing because a cubic packing of 14 spheres can be distinguished in the layered sphere structure – the basic cube can be regarded as a reflection of the entire structure.

Material: Cellulose spheres (white, 30 mm), glue.

Procedure: Glue together a triangle "with central hole" from 27 spheres (see picture below). Layer spheres as tightly as possible. Determine the coordination number. Construct the basic cube (see picture) using 14 spheres. Place the cube totally into the packing.

Observation: A regular tetrahedron is formed by the spheres, the coordination number is 12. It is possible to completely fit the basic cube into the tetrahedron packing with the use of the "central hole" for one corner of the cube.

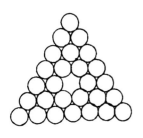

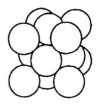

E4.6 Solution of Iodine in Ethanol

Problem: Both iodine and ethanol consist of molecules – one can therefore declare both iodine particles and ethanol particles for using the particle model of matter. In order to make the dilution experiment more interesting, one should take iodine crystals in a pipette and demonstrate the larger density of the solution in comparison to the solvent. One can further show the solution of iodine in gasoline and demonstrate the violet color of this solution, however gasoline is a mixture of different hydrocarbons, making the interpretation with particle model more complex.

Material: Gas jar or curette, pipette, forceps; iodine, ethanol.

Procedure: Fill a gas jar with ethanol, fix a pipette half-dipped into the ethanol. Take a crystal of iodine and place it in the pipette (see Fig. 4.5).

Observation: A fine stream of brown liquid flows from the pipette and spreads over the bottom of the container. After a while the entire liquid is colored light brown.

E4.7 Sphere-Model for the Solution of Iodine in Ethanol

Problem: The particle model of matter may be introduced through the solution of two substances (see E4.6). Students will not come up independently with the particle interpretation – they need help by the verbal particle explanation or better by demonstrating model experiments. One can use one big sphere as a model for the iodine particle and one little sphere as a model for the ethanol particle. The model experiment can be done without many comments – one could then ask the students to associate certain aspects of the model with the original (see E4.6). In the model discussion it is important to expose the model traits, but also the irrelevant items. One can compare the model experiment with the model drawing (see Fig. 4.6).

Material: Large glass bowl (on the overhead projector); big and small cellulose spheres (i.e. 30 mm and 12 mm diameters respectively).

Procedure: Fill the surface of the large glass bowl up to three quarters with small spheres (models for ethanol particles). Create some space in the middle and place 16 or 25 big spheres arranged in a square shape (models for iodine particles in iodine crystal). Carefully shake the glass bowl until all spheres are mixed and have spread over the bowl.

Observation: At first the special arrangement of big spheres is observed, after shaking the bowl all moving spheres are spread uniformly over the entire surface of the bowl.

E4.8 Reaction of White and Red Phosphorus with Oxygen

Problem: The P atom is neither white nor red, only after many P atoms have been arranged in a crystal that can be determined as white or red phosphorus. If small portions of both phosphor modifications are ignited to react in oxygen, one attains in both cases the same white smoke of phosphorus oxide, which always produces the same acidic solution when diluted in water: phosphoric acid solution. This oxygen reaction therefore makes it plausible that there are not two types of P atoms, but just one type. P atoms may be linked together in two different ways: one way leads to P_4 molecules and to the chemical structure of white phosphorus, the other way to the infitie structure of P atoms in red phosphorus.

Material: 2 gas jars, 2 glass plates, combustion spoon; white and red phosphorus, water, oxygen (lecture bottle), universal indicator paper.

Procedure: Fill two gas jars with oxygen by replacement of air. Give a small white phosphorus crystal in the combustion spoon, ignite with the burner and place in the jar. Repeat the reaction with red phosphorus. Fill some water in both cylinders, shake and examine the solution using indicator paper.

Observation: Both phosphorus types react with oxygen generating bright light and a dense white smoke, which in part covers the inside wall of the cylinder. The smoke can be absorbed in water, the resulting solution colors the indicator paper red.

Disposal: The remains of the phosphorus can be completely burnt under the fume hood with the aid of the burner flame or can be placed in concentrated copper sulfate solution.

E4.9 Volumes of Liquid and Gaseous Ethanol

Problem: The dramatic increase in volume through evaporation of a liquid can be explained through the much faster movement of the smallest particles and the resulting greater distance of particles from each other. The free distance of molecules in a gas increases with temperature. Children tend to be captivated by the idea of the "horror vacui" and believe that the particles grow bigger and thereby fill the greater volume. In order to convince them that the particle size does not change, the vapor is further condensed and then one can show the students that the same volume of ethanol is present. The discussion might convince students that the space between the gas particles is free of matter.

Material: Beaker (1000 ml), tripod and wire gauze, burner, balloon; ethanol.

Procedure: Fill a balloon with 2 ml ethanol and close tightly with a knot. Boil 500 ml water in a beaker, insert the balloon in the boiling water. Remove the balloon after some time.

Observation: The balloon is blown up to its full size and retains its size in the boiling water. If it is removed from the water it shrinks and returns to its original size. The same little amount of ethanol remains in the balloon.

E4.10 Model Experiments for the Three States of Matter

Problem: The idea behind this experiment is to compare the big volume of a gas portion versus the same mass of the liquid or the solid. The increase in volume is shown by strongly shaking the chosen amount of spheres in a large glass bowl. It is however necessary to point out in the model discussion, that in nature there is nobody to do the shaking, the small particles move themselves and independently. Apart from this, in such a model one finds air between the spheres, in the original substance there is nothing between the particles.

Material: Watch glass, large glass bowl, identically-sized small spheres.

Procedure: Place 15 spheres on a watch glass, then project them with the help of the overhead projector. Slightly move the watch glass manually without moving spheres from their places. Toss the spheres into the glass bowl and move the bowl slightly, so that spheres remain together but change their places. Then move the glass bowl on the projector quickly back and forth, so that the spheres separate from each other and spread over the entire bowl.

Observation: The spheres remain in place on the watch glass, they demonstrate in the model how particles are arranged in certain formations and thereby form certain crystal patterns. If they are slightly shaken back and forth on the watch glass, the spheres even symbolize the vibration of particles in a crystal at room temperature. In the second model a liquid is shown: the spheres touch each other but they change their places through the constant movement. In the third model the vapor is visualized: the distances between the spheres are much larger because of the fast movement, the spheres fill out the entire space in the glass bowl.

E4.11 An Empty Flask is Full of Air

Problem: The term "empty flask" may mean to the naked eye that neither a liquid nor a solid are present, but we know on the other hand that the flask is really filled with air. So one can pump out the entire air from a flask: in this case the flask contains "nothing". In order to reflect upon the "empty flask", the following small experiments are carried out.

Material: Beaker (200 ml), large glass bowl, syringe with stop cock, glass tube.

Procedure: Fill a glass bowl with water and place a beaker vertically upside down inside it. Tilt the beaker, fill with water, and hold tightly. Fill a syringe with 100 ml of air and transfer the air into the water-filled beaker. The air can again be removed from the beaker though sucking it out with the help of the empty syringe and a glass tube.

Observation: The water remains in the beaker. The water-filled beaker takes 100 ml of air from the syringe, then the air can be sucked back into the syringe again: air is a substance with a specific density (see E3.13), a special mass of air takes a special volume at normal pressure.

E4.12 A Flask Contains Nothing

Problem: Because it's impossible to imagine of "nothing", it is necessary to demonstrate the vacuum though experimental experiences. So students may observe how all air is removed out of a glass container: it contains neither air nor other matter, it contains nothing. Based upon these simple experiments, the statements "a flask is empty" and "a flaks contains nothing" can be reflected upon, and associated differences can be discussed. The term "nothing" can be especially used for the particle model and makes plausible that there is "nothing" between the gas particles – not even air!

Material: Round flask with outlet, stopper and stop cock, syringe, water jet-pump (water aspirator), large glass bowl.

Procedure: Remove 50 ml of air from a round flask by sucking it out with the help of a syringe, then close the flask by the stop cock. Once again attempt to remove 50 ml air. When it becomes too difficult to further remove air with the syringe, then attach a water-jet pump and let the water run for a minute. Dip the outlet of the closed flask into the glass bowl under water and open the outlet.

Observation: A certain amount of air can be removed from the round flask through the use of the syringe, low air pressure is thereby created. After the rest of the air is removed using the water-jet pump, a vacuum results in the flask: no air or other material is left in the flask. If one opens the outlet of the flask under water, it fills completely with water.

References

1. Engida, T., Sileshi, Y.: Concept Cartoons as a Strategy in Learning, Teaching and Assessment Chemistry. Addis Ababa, Ethiopia, 2004
2. Stachowiak, H.: Gedanken zu einer allgemeinen Theorie der Modelle. Studium Generale 18 (1965), 432
3. Steinbuch, K.: Denken in Modellen. In: Schaefer, G., u.a.: Denken in Modellen. Braunschweig 1977 (Westermann)
4. Barke, H.-D.: Modelle, Modellvorstellungen. In: Barke, H.-D., Harsch, G.: Chemiedidaktik Heute. Lernprozesse in Theorie und Praxis. Heidelberg 2001 (Springer)
5. Asselborn, W., Jaeckel, M., Risch, K.T.: Chemie Heute. Hannover 2001 (Schroedel)
6. Pfundt, H.: Das Atom – Letztes Teilungsstueck oder Erster Aufbaustein. Chimdid 7 (1981), 7
7. Barke, H.-D.: Der "horror vacui" in den Vorstellungen zum Teilchenkonzept. In: Barke, H.-D., Harsch, G.: Chemiedidaktik Heute. Lernprozesse in Theorie und Praxis. Heidelberg 2001
8. Novick, S., Nussbaum, J.: Pupils' understanding of the particulate nature of matter: A cross-age-study. Science Education 85 (1981), 187
9. Kircher, E., Heinrich, P.: Eine empirische Untersuchung über Atomvorstellungen bei Hauptschuelern im 8. und 9. Schuljahr. Chimdid. 10 (1984), 199
10. Fladt, R.: Kurskorrektur im Chemieunterricht. Dargestellt an der Einfuehrung und Anwendung des Teilchenmodells in der Sekundarstufe I. MNU 37 (1984), 354
11. Weninger, J.: Didaktische und semantische Probleme bei der Einfuehrung der Atomhypothese und der Kern-Elektron-Hypothese. In: Weninger, J., Bruenger, H.: Atommodelle im naturwissenschaftlichen Unterricht. Weinheim 1976 (Beltz)

12. Buck, P.: Wie kann man die ‚Andersartigkeit der Atome' lehren? Chem. Sch. 41 (1994), 460
13. Barke, H.-D.: Raumvorstellung zur Struktur von Teilchenverbaenden. In: Barke, H.-D., Harsch, G.: Chemiedidaktik Heute. Heidelberg 2001 (Springer)
14. Thurstone, L.L.: Primary mental abilities. Psychometric Monographs 1 (1938)
15. Bloom, B.S.: Stabilität und Veränderung menschlicher Merkmale. Basel 1971 (Beltz)
16. Engida, T.: Structural Chemistry and Spatial Ability in Chemical Education. A Case of selected German and Ethiopian Schools. Dissertation. Muenster 2000
17. Rost, D.: Raumvorstellung. Psychologische und paedagogische Aspekte. Weinheim 1977 (Beltz)
18. Harsch, G.: Stereobilder zum Training des Raumvorstellungsvermoegens. In: Barke, H.-D., Harsch, G.: Chemiedidaktik Heute. Heidelberg 2001 (Springer)
19. Möller, B.: Chemische Strukturen entdecken und verstehen. Entwicklung und Erprobung einer multimedialen Lernhilfe. Dissertation. Muenster 2004 (Schueling)
20. Sopandi, W.: Raumvorstellungsvermoegen und Chemieverstaendnis im Chemieunterricht. Dissertation. Muenster 2005 (Schueling)
21. Maccoby, E.E., Jacklin, C.N.: The Psychology of Sex Differences. Stanford 1974 (University Press)
22. Brockmann, K.: Das Teilchenmodell im Anfangsunterricht. Empirische Erhebung diesbezueglicher Modellvorstellungen bzw. Fehlvorstellungen und Vorschlaege zu deren Korrektur. Staatsexamensarbeit 2006 (Universitaet Muenster)

Further Reading

Barker, V.: Chemical concepts: Particles are made of this. Education in Chemistry 38 (2001), 36
Barker, V.: Chemical concepts: Changing matter. Education in Chemistry 38 (2001), 92
Benson, D.L., Wittrock, M.C., Baur, M.E.: Students' preconceptions of the nature of gases. Journal of Research in Science Teaching 30 (1993), 587
Brook, A., Briggs, H., Driver, R.: Aspects of Secondary Students Understanding of the Particle Nature of Matter. Leeds 1984 (University of Leeds)
Gabel, D.L., Samuel, K.V., Hunn, D.: Understading the particulate nature of matter. Journal of Chemical Education 64 (1987), 695
Griffiths, A.K., Preston, K.R.: Grade-12 students' misconceptions relating to fundamental characteristics of atoms and molecules. Journal of Research in Science Teaching 29 (1992), 611
Haidar, A.H., Abraham, M.R.: A comparison of applied and theoretical knowledge of concepts based on the particle nature of matter. Journal of Research in Science Teaching 29 (1991), 611
Johnson, P.M.: Progression in children's understanding of a 'basic' particle theory: A longitudinal study. International Journal of Science Education 20 (1998), 293
Jones, B.L., Lynch, P.P., Reesink, C.: Children's understanding of the notions of solid and liquid in relation to some common substances. International Journal of Science Education 11 (1989), 417
Lee, O., Eichinger, D.C., Anderson, C.W., Berkheimer, G.D., Blakeslee, T.D.: Changing middle school students' conceptions of matter and molecules. Journal of Research in Science Teaching 30 (1993), 249
Mitchell, A.C., Kellington, S.H.: Learning difficulties associated with particulate theory of matter in the Scottish Integrated Science Course. European Journal of Science Education 4 (1982), 429

Nowick, S., Nussbaum, J.: Junior high school pupils' understanding of the particulate nature of matter: An interview study. Science Education 62 (1978), 273

Osborne, R.J., Cosgrove, M.M.: Children's conceptions of the changes of state of water. Journal of Research in Science Teaching 20 (1983), 825

Stavy, R.: Children's conceptions of changes of matter: From liquid (or solid) to gas. Journal of Research in Science Teaching 27 (1990), 247

Fig. 5.1 Concept cartoon concerning ionic bonding

Chapter 5
Structure–Property Relationships

Initial structure–property relationships have been studied using examples like "diamond/graphite" and "white/red phosphorus" (see Sect. 4.3), the modifications have been established from the same C atoms or P atoms respectively. However, the substances are drastically different in their chemical structure and therefore in their characteristic properties. The misconceptions could be corrected with the consideration that the individual C atom or P atom show absolutely no properties like color or density. Such characteristics can be determined only when a little crystal is visible (see Sect. 4.3).

It is much easier to explain the concept of structure–property relationship when one combines both **chemical structure** and **chemical bonding** in the discussion. One can then refer back to volatility, solubility, melting or boiling behavior of materials in relation to the question of bonding between atoms, ions and molecules. Connections between structure and properties also play a big role in the following themes: homologous series, functional groups, stereochemistry, dyes, synthetics, carbohydrates, fats, proteins, etc. It is not, at first, necessary to take these familiar themes from organic chemistry into consideration because the discussion requires the electron-pair bonding and also the wave-particle duality of the electron that naturally cannot be understood, and it is not possible to avoid misconceptions.

It would be better to produce examples regarding the structure–property relationships that do not require chemical bonding models, but can be answered using **chemical structures** (see Fig. 2.3). Examples of this would be the structures of metals and alloys with their specific arrangements of atoms and the structures of salts and those arrangements of ions.

5.1 Structure and Properties of Metals and Alloys

"Iron atoms can rust and turn red-brown in color; iron atoms are hard, lead atoms are soft; bass atoms are shiny-gold".

This conflict between properties of a substance on the one hand and characteristics of a single atom on the other hand is easily formed because arrangements of many atoms are so seldom used in chemistry lessons for explaining the properties of substances, especially the properties of metals and alloys. As long as one continues to examine and discuss the single atom, many students stick with such ideas: "the gold atom is yellow".

These facts have already been discussed in the examples of diamond/graphite and red/white phosphorus (see Sect. 4.3). The varying properties can be shown by differences in chemical structures. However, these structures are not easy to understand. Because it is possible to correctly demonstrate the arrangement of metal atoms using closest-sphere packing models, it is useful to look at metal structures with regard to the property–structure relationships and try to address the above-mentioned misconceptions.

Teaching and Learning Suggestions. It is not possible to find such aesthetically pleasing metal crystals compared to salt crystals, for instance nicely formed alum crystals (see Fig. 4.4). Metal crystals can be obtained through precipitation from saline solutions (E4.4) or easily prepared by electrolytic deposition (see E5.1). Very impressive trees of silver crystals can be formed from a silver salt solution (see Fig. 5.2).

The arrangement of Ag atoms in silver crystals can be described as cubic closest packing of spheres in space: each sphere corresponds to an Ag atom in the packing model. The study of closest sphere packings leads to two different stacking sequences (see E5.2): (1) triangular layers in the sequence of ABCABC, and (2) in the sequence of ABABAB can be stacked (see Fig. 5.3). The coordination number 12 can be seen in both of the closest packings (see Fig. 5.3): each sphere is surrounded by six spheres in one layer, plus three from above, and three from below (see E5.2). Both stacking sequences lead to pyramidal shapes, in every one basic hexagonal elements can be detected (see Fig. 5.4).

Fig. 5.2 Silver crystals through electrolysis [1]

5.1 Structure and Properties of Metals and Alloys

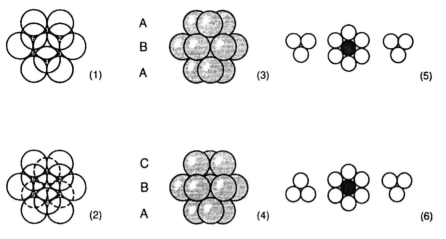

Fig. 5.3 Layer sequencing ABCA and ABA of two types of closest sphere packings [2]

However, in the regular pyramid (see (2) in Fig. 5.4) one finds an additional cubic element which is made up of 14 spheres (see Fig. 5.5). This cube can be mounted into the pyramid of the stacking sequence ABCA standing on one of the eight corners of the cube (see E5.3). For this reason, this packing of spheres is known as the **cubic closest packing.** It can also be formed from layers with spheres arranged in squares, here the elementary cube is standing on one of the six faces (see E5.3).

Because each face of the basic cube contains one sphere in the face center, experts know the structure as the **cubic face centered structure.** The expert knows the unit cell as well as the elementary cube (see Fig. 5.5), which is achieved through mentally cutting the spheres in the elementary cube through

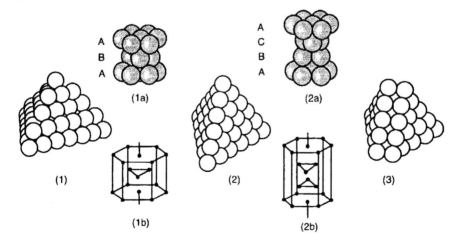

Fig. 5.4 Two types of closest sphere packings [2]

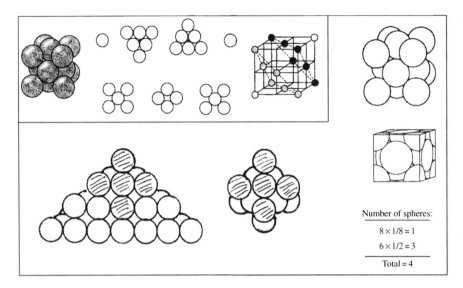

Fig. 5.5 Elementary cube and unit cell of the cubic closest structure [2]

the center of each sphere parallel to the surface of the cube. It is suitable, for demonstration purposes, to show infinite crystal structures by moving the unit cell in all three spatial directions. If one joins the spherical parts of the unit cell, one gets four complete spheres: one can count six half-spheres on the cubes faces and eight one-eighth spheres on each of the eight corners (see Fig. 5.5). For demonstration purposes, a unit cell can be built with cardboard (see E5.3). Wirbs [3] was able to establish that the comprehension for unit cells can be achieved and cubic unit cells can be successfully interpreted in school grades 8 and 9 classes.

The cubic closed ABC type packing (ccp) or cubic face centered structure is realized in nature by metals like gold, silver, copper, lead, aluminum, nickel and many other metals: based on the importance of the metal copper, it is called the **Copper Type** (see Fig. 5.6).

The densest ABAB close-packing structure is also known as the hexagonal closest packing structure (hcp) because of its hexagonal elementary unit. It can be found in metals like magnesium and zinc: **Magnesium Type** (see Fig. 5.6).

A third metal structure is known as cubic body centered (cbc), i.e. eight spheres touch a sphere in the middle of an imagined cube. This metal structure is not a dense close-packing structure – it can be found in alkali metals and in tungsten: **Tungsten Type** (see Fig. 5.6).

The effects of the various metal structures on their properties can be observed in their **ductility.** If one hammers a copper plate and a magnesium ribbon for quite some time (see E5.4) one discovers that copper plate can be hammered down to a very thin foil (similar to gold metal leaf). In comparison, the magnesium ribbon crumbles into lots of small pieces – there is no such thing as a magnesium leaf!

5.1 Structure and Properties of Metals and Alloys

Lattice structure	Coordination number	Structure model	Example
Hexagonal	12	hexagonal close packing	Magnesium, Zinc: **Mg-Type** 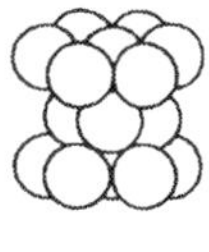
Face-centered cubic	12	cubic close packing	Copper, Silver, Gold, Biel: **Cu-Type** 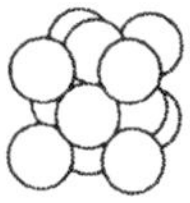
Body-centered cubic	8	body centered packing	Alkali metals, Tungsten: **W-Type** 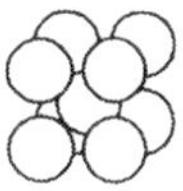

Fig. 5.6 Crystal structures of metals, symbols and examples [2]

In the case of the cubic closed structured metal crystal, there are glide planes within the atomic layers in all directions, thus the mechanical attack can be repelled from many directions through the movement between atomic layers. The elementary cube itself allows the movement of smooth triangular layers in four directions, i.e. perpendicular to the four diagonal spaces. In comparison, the hexagonally closed structured metal crystal has only one direction of the triangular layer in the hexagonal unit cell. These metals are not very ductile and crumble to dust when force is applied.

Blacksmiths and sword makers throughout time had similar experiences, i.e. iron has to be red hot before it can successfully be forged. It becomes harder as soon as it cools off. The iron structure is cubic body centered at room temperature and changes into the cubic face centered structure without diffusion when it reaches the high temperature of 910°C. This so-called γ-iron can be forged like copper and noble metals; after cooling it down to α-iron, the good ductility is lost, the metal is hard again [2].

If one wants to make the iron even "harder", then one should use an iron-carbon alloy that contains up to 2% carbon: steel. It is much harder because of the statistical integration of C atoms into interstices of the lattice, which causes further blocking of the glide planes in the lattice. The same happens when a silver-tin mixture is "amalgamated" with mercury, the classical material used by dentists for filling teeth. The alloy is much harder than any of the original metal substances. The same applies to amalgamated sodium that is formed by reaction of sodium with mercury (see E5.5).

The **structure of alloys** and structural changes at certain temperatures can explain the memory effect of special alloys (see E5.6) like "Nitinol", a nickel-titanium alloy [4]. The structure as well as the much greater electric conductivity explains the difference between the copper-gold structures (CuAu and Cu$_3$Au) and compounds of "red-gold" alloys [2]. Further examples of alloys can be used in the discussion and explanations of properties with the help of chemical structures.

5.2 Existence of Ions and Structure of Salts

"The water molecules evaporate, the salt molecules remain; salt can be found at the bottom in the form of NaCl particles; all that remains are sodium chloride particles" [5].

While the terms atom and molecule have been introduced at an early stage by most teachers and have quickly been integrated by the students in their vocabulary, teachers and students often have difficulties with the term "ion". Ions are introduced at a much later stage in association of the ionic bonding, but not in the sense of already existing smallest particles of salts. Several empirical studies have demonstrated this [5].

Ions in Precipitation Reactions. Grade 10 students of German academic high schools have learnt the atomic model and about the idea of the ion and ionic bonding in their chemistry lessons. These students saw precipitates of calcium sulfate from saturated salt solutions and have been asked to imagine the smallest particles in these solutions before and after the precipitation [6]. The expected ion symbols of the initial solution were correctly supplied in 50% of the cases. However, the other half of the student group has shown misconceptions of "salt molecules" or of "electron transition in the formation of ions from atoms". With respect to the precipitation product, only 30% of the students provided acceptable structural models, the amount of misconceptions grew to 70% [6].

Ions in Salt Solution. Students at the upper grades were given symbols representing ions in a salt solution (see Fig. 5.7, "before evaporation"). Afterwards, they were asked to describe what happens to the ions when the water evaporates [7]. Apart from several correct answers regarding the ions by crystallization of sodium chloride, a large percentage of answers were given based on the existence of NaCl

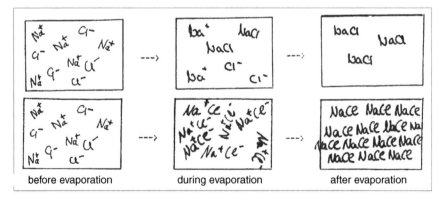

Fig. 5.7 Mental models of a table salt solution: before, during and after evaporation [7]

5.2 Existence of Ions and Structure of Salts

molecules in crystals. These students started with ions in the solution, however when developing mental models for the evaporation of water they argued the "neutralization" of ions [7] and the continuous fusion of ions into molecules and finally they drew NaCl molecules as particles of the crystal (see Fig. 5.7).

Ions in Mineral Water. In a questionnaire regarding the label on a bottle of mineral water, students in upper grades were shown the name of the substances contained in the mineral water: calcium chloride, magnesium chloride, sodium chloride and sodium bicarbonate [8]. The point of the questionnaire was to test their knowledge of the term ion and to see if they could successfully apply it to saline solutions. In order to introduce the correct ion term into the discussion and to suggest misconceptions regarding "salt molecules", specific alternatives were given: "mineral water contains salt molecules" and "mineral water contains ions of different salts" (see Fig. 5.8). Students were encouraged to suggest their own answers too.

The decisive tasks were students' drawings regarding their mental models. First they were required to select the most fitting alternative. After choosing "ions of different salts" they should produce model drawings to demonstrate the types of ions in that mineral water. 132 students, grade 9–12, took part in the written test [8].

Despite the fact that all students had dealt with the ion term in class, only 25% of the students recognized "ions of different salts" as the correct answer. Approximately the same amount chose "salt molecules". If one looks at the model drawings, a mere 4% of students actually included ion symbols in their drawings. Many of the test persons who, although they crossed off the ions as the correct answer, chose symbols for molecules (see Fig. 5.8); their mental models have been NaCl or Na-Cl, $CaCl_2$ or Cl-Ca-Cl, etc.

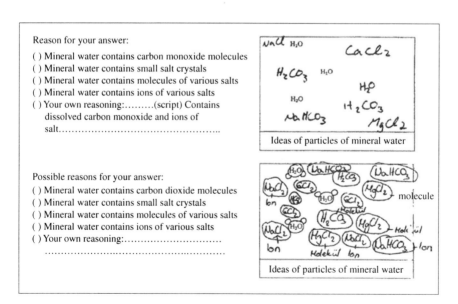

Fig. 5.8 Examples for misconceptions regarding particles in mineral water [8]

Additional model drawings showed the following misconceptions: (a) ion symbols were, as in the structure symbols for the molecules, joined together by hyphens, (b) metal atoms and non-metal atoms were combined, (c) using the atomic shell model, a "transfer of electrons" from one atom to another was sketched (ions are always "created" in the process), and (d) similar spheres were drawn in and marked with "NaCl": "sodium chloride particles" or "salt particles" as from the simple particle model of matter. In this regard, it is particularly clear that the misconceptions are "school-made". The students enter the class without any knowledge of the ion term and definitely learn it from the teacher or from the textbook. It seems that there are huge instructional difficulties in teaching the ion concept in a comprehensible way.

It appears as if these difficulties exist everywhere, even in St. Petersburg, Russia. Viktor Davydow from the Herzen Pedagogical University, hypothesized that the Russian students would be better in answering the questionnaire. The aforementioned questionnaire was translated and administered to students of grade 9 and 10. They also visualized molecules and partially marked them with the Lewis-notation, presented ion symbols with "bonding lines" like in symbols of molecules, or described both: molecules and ions (see Fig. 5.9).

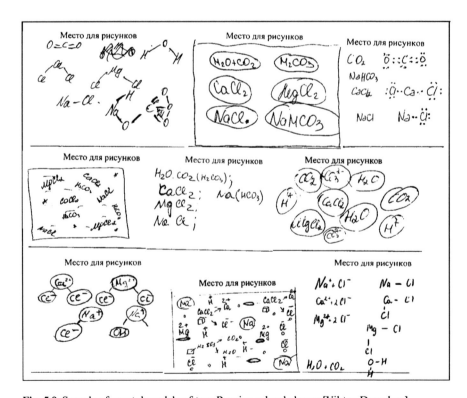

Fig. 5.9 Sample of mental models of two Russian school classes [Viktor Davydow]

5.2 Existence of Ions and Structure of Salts

Teaching and Learning Suggestions. The depicted empirical experiences show, that after lessons about the differentiated atomic model and ionic bonding, the existence of ions as the smallest particles of salts is only inadequately anchored in the young people's minds. Ion symbols cannot be satisfactorily applied by the majority of the test takers. Undifferentiated symbols like NaCl and HCl in reaction equations are mostly used and the learners are left to their own devices how these compounds are structured: by molecules or by ions [6].

These experiences have been described in literature of chemistry education for decades [6]. Only in exceptional cases has the teaching of chemistry been completely altered due to this criticism [9]. Students are not only given the atoms and correlative atomic masses based on the Daltonic model with the Periodic System of the Elements (PSE), but also important ion types (see Fig. 5.10). This special Periodic System graphically depicts and clearly arranges "atoms *and Ions* as basic particles of matter" [9] (see Fig. 5.10). Both spherical models of atoms and ions correlate to their sizes as measured by physicists. Christen [10] uses a similar system in his book for introduction to chemistry.

With this information one is informed about the ion charge due to the position in the PSE, but the balance of protons in the atomic core and electrons in the atomic shell is not intended at this time; this balance will be introduced at a later stage of the chemistry course. To aid in the orientation and combination, the basic particles of matter are separated in "metal atoms" on the left-hand side

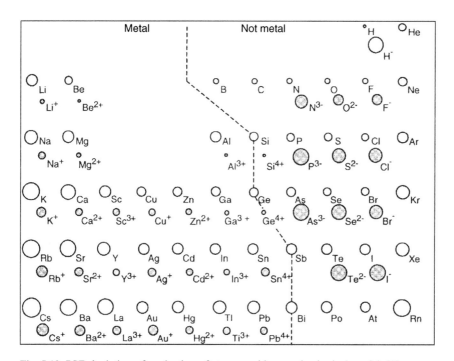

Fig. 5.10 PSE-depiction of a selection of atoms and ions and spherical models [9]

of the PSE and in "non-metal atoms" on the right-hand side: therefore the H-atom and the hydride ion is unusually but consciously placed on the right-hand side of the PSE (see Fig. 5.10). Because the H^+ ion or proton does not possess an independently existing particle for an ion lattice but rather engrosses the electron cloud of a molecule (see Chap. 7), it is not shown in this Periodic Table.

Should one need to substantiate the existence of ions, certain measurements such as the lowering of the freezing point can be carried out [11]. If one takes each 1 molar solution of ethanol, sodium chloride and calcium chloride and determines the temperature at which the solutions begin to solidify (see E5.7), in an ideal case, one measures $-1,9°C$, $-3,8°C$ and $-5,7°C$, respectively. Because freezing-point temperatures are not dependent on the type but rather on the number of particles per volume unit [11], it is concluded that

- 1 mol ethanol → 1 mol particles → 1 mol C_2H_5OH molecules,

- 1 mol NaCl → 2 mol particles → 1 mol Na^+ ions + 1 mol Cl^- ions,

- 1 mol $CaCl_2$ → 3 mol particles → 1 mol Ca^{2+} ions + 2 mol Cl^- ions.

One can begin the lesson with the PSE-collection of atoms and ions as basic particles of matter (see Fig. 5.10). Metal atoms can be mentally combined to metal crystals and crystals of alloys, and can be illustrated through closest sphere packings. This way of teaching chemistry by mentally combining "atoms and ions as basic particles of all matter" may be called "structure-oriented approach" or "structure-oriented chemistry". Wirbs [3] was able to empirically prove that the special Periodic Table and the structure-oriented approach was accepted by her students as a "central thread" in chemistry lessons. Her grade 8 students were able to mentally combine metal atoms "left and left in PSE"; to understand sphere packing models; and finally to successfully add up the content of unit cells and derive formulas like CuAu and Cu_3Au [3].

It is also possible to derive factual definitions for the terms "element and compound" using the collection of basic particles of matter. In an element, atoms of one atom type are combined; in a compound, there are at least two types of atoms or ions. All salts are included in this sentence as being compounds of at least two kinds of ions – other incorrect mental models can be avoided: "salt particles", "atoms", "formula units", or "compound units" [6].

After combining the metal atoms, ions can be combined "left and right in PSE", perhaps model-like and exemplary for simple structures of salt crystals. Due to the ion charge and the balancing of the specific electrical charges for an ionic compound, formulas like $(Na^+)_1(Cl^-)_1$ or NaCl and $(Mg^{2+})_1(Cl^-)_2$ or $MgCl_2$ could be established (see Fig. 5.11).

Strehle [12] was able to show, using structure-oriented chemistry lessons in a grade 9 class at a school near Münster in Germany, that the mental combination of ions "left and right in PSE" caused no difficulties for these students. She

5.2 Existence of Ions and Structure of Salts

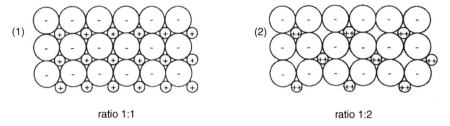

ratio 1:1   ratio 1:2

Fig. 5.11 2-D Model drawings for arrangements of ions in salt crystals [2]

was also able to prove that the drawing of two-dimensional models and the building of spatial sphere packings led to a good understanding regarding ions in ion lattices of salts. She could further show that the ion concept was anchored in the students' minds for long-term memory.

One can also find the correct ratio for salt solutions and can discover through the relating model drawings, that ions are present by independently from each other. If one wishes to bring up hydrated ions, one should additionally choose the well-known symbol (aq). Should precipitation reactions be demonstrated, they could be interpreted by showing the related ions, then one could use shortened model drawings as a didactic aid (see Fig. 5.12). Chemical equations can be easily derived with the help of these model drawings, especially those equations solely symbolizing the ions which form solid matter (see Fig. 5.12, last row).

In order to arrive at the important **sodium chloride structure**, one could look at and discuss the interstices in the cubic closest sphere packing model [2]: interstices that are octahedral and tetrahedral can be found (see E5.8). If one fills the octahedral interstices with smaller spheres, one ends up with the sodium chloride structure (see Fig. 5.13). The larger spheres symbolize the chloride ions, the smaller spheres the sodium ions. The structure can be described as the cubic closest packing of chloride ions, whose octahedral interstices are completely filled by sodium ions. For other salt structures, only part of the octahedral sites are filled, as in aluminum oxide where the ratio of ions 2:3 applies [2].

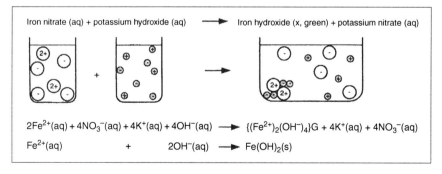

Fig. 5.12 Model drawings for a precipitation reaction (see also Fig. 5.23) [2]

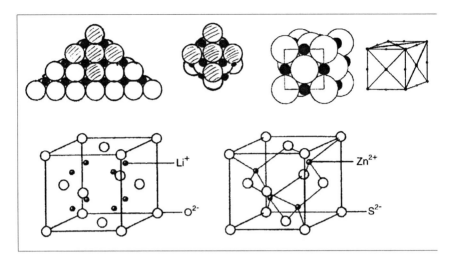

Fig. 5.13 Structural models for cubic salt structures like NaCl, ZnS and Li$_2$O [2]

The coordination numbers for both ion types in the NaCl structure are 6; meaning one Na$^+$ ion is surrounded by 6 Cl$^-$ ions, and one Cl$^-$ ion by 6 Na$^+$ ions (see E5.8). These numbers should also be identified in the elementary cube of the NaCl sphere packing (see Fig. 5.13). Also unit cells and lattices should be derived from the elementary cube (see E5.8). In this way, several models for the sodium chloride structure can be observed. They can be compared and irrelevant ingredients can be discussed. Thereby, appropriate mental models can be created in the student's mind.

Many other salt structures could be described by using various fillings for octahedron or tetrahedron interstices [2]. As examples for the allocation of tetrahedron site, the zinc sulfide structure and the lithium oxide structure have been included (see Fig. 5.13).

The formation of ions in salt crystals carries certain consequences for their properties: if one hammers on a metal block, it will be shaped according to the strength of the blows; if a certain force strikes a salt crystal, it splits or breaks off into various pieces (see E5.9). The explanation lies in the nature of the repulsion of equally charged ions. If the anions and the cations, which are positioned opposite each other, are forced to come close to each other, repulsion sets in which overcomes the attractive force of differently charged ions (see Fig. 5.14).

Even the electrical conductivity results from the presence of mobile ions. A rock salt crystal does not at first conduct electricity at room temperature; however, the strong heating of the crystal leads to an ever-increasing conductivity (see E5.10). If a salt is melted or diluted in water, a good electrical flow can be attained (see E5.10). Using direct current voltage leads to electrolysis and to electrolytic decomposition of the salt melt or the salt solution.

5.3 Mental Models on Ionic Bonding

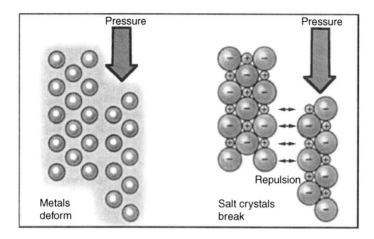

Fig. 5.14 Structure–property relationship of metal crystals and salt crystals

Evaporation of the water from salt solutions results in solid salt crystals: the ions involved form an ion lattice corresponding to the salt structure. If one allows the water to slowly evaporate from the saturated solution, this often results in large and beautiful crystals. Particularly the alum salt, when growing crystals from saturated solutions (E4.1), results in beautifully formed octahedron (see Fig. 4.4). In the process, K^+(aq) ions, Al^{3+}(aq) ions and SO_4^{2-} ions join together to form an ion lattice of cubic symmetry. If one adds approximately 10% of urea to a saturated sodium chloride solution the salt crystals do not crystallize in the expected cubic form, but in an octahedron form with identical symmetry elements as the cube.

If one melts a particular salt like sodium acetate and allows the molten matter to cool, it amazingly remains molten, even at room temperature, and does not crystallize. If one adds a little sodium acetate crystal into this molten mass, one instantly gets white salt crystals; the heat of crystallization is immediately released, and the temperature is about 50°C (see E5.11). This reaction is used to produce so called "pocket warmers". Following the crystallization, they are heated again to form the molten salt; after cooling them to room temperature, they will deliver heat again. Precipitation reactions are also suitable for demonstrating the crystallization of certain types of ions (see E5.12).

5.3 Mental Models on Ionic Bonding

The arrangement of ions in a solid crystal can be demonstrated in two-dimensional models (see Fig. 5.11). The ionic bonding describes a special stability of electrostatic attractive forces of differently charged ions, and of electrostatic repulsion forces of similarly charged ions. Students usually do not possess such or similar explanations regarding electrostatic forces.

Empirical research regarding ionic bonding shows other concepts. Based on the information supplied in lessons relating to the atomic model with nucleus and electron shells, one mostly comes up with concepts of imaginary momentary transfer of electrons in a redox reaction, with ionic *bonding* through the *formation* of ions, or with simple bonding between individual ions or with ion pairs. Even salt molecules and salt particles exist in the minds of students. Hilbing [13] incorporated these findings into a cartoon (see Fig. 5.15). Schwoeppe [14] further researched similar misconceptions and suggested specific lessons to remedy them. Both works have been compiled and supplemented.

Ionic Bonding and Salt Particles. In his examinations, Hilbing [13] asked a group of German gymnasium students of grade 10 about ionic bonding and how they describe the solid salt after evaporation of water from salt solution. Most of the participants drew pictures or answered with "salt particles" or "NaCl particles". It appears to be natural to rely upon the old tried and trusted concept of particles.

Even after teaching the traditional lesson on ionic bonding, a high school teacher came across these concepts of salt particles or NaCl particles again: "although the students have learned a lot about ions and ionic charges over a period of 10 lessons, they still produce concepts which are comparable to the status quo before any of these lessons were introduced. That means that this concept does not outlast the lesson. It does *not* mean that the students do not know the 'correctly' intended concept taught in the lesson unit but rather that

Fig. 5.15 Empirical findings towards student's misconceptions of ionic bonding [13]

5.3 Mental Models on Ionic Bonding

they are not completely convinced by this new mental model: they state the familiar concept of 'salt particles' or 'NaCl particles' – just to be on the safe side" [13].

"Such concepts are known as lower cognitive layers according to Niedderer. It means that students have both a concept of salt-particles as basic particles of sodium chloride, as well as Na^+ ions and Cl^- ions as the smallest particles: competitively, they mix up both images so often. Regarding our question, the students have activated the cognitive layer of the particle model but not the intended concept of ions and ionic bonding, which had been taught by the teacher. One must expect that the alternative concept of salt particles later supplies a point of connection for new contents i.e. in topics of acids and bases: Who is not aware of H-Cl molecules in hydrochloric acid, of Na-O-H molecules in sodium hydroxide solution?" [5].

Ionic Bonding and Electron Transfer. According to Taber [15], students may understand the origin of ions and atoms through electron transfer, but often tend to confuse ionic bonding with this electron transfer and claim that electron transfer is the same as the concept of ionic bonding. The model drawing of a student regarding ionic bonding (see Fig. 5.16) provides an example of this mental model: two ions result from electron transfer of two relating atoms. Students mix electron transfer with ionic bonding: in the drawing the student wrote "ionic bonding" for the image of an electron transfer (see Fig. 5.16).

The students do not differentiate between the formation of ions and ionic bonding and therefore develop misconceptions regarding ionic bonding. Taber determines that about 58% of students he questioned were of the opinion that ionic bonding only came about because the atoms "retain a complete outer shell" – as a result, they have a concept regarding ionic bonding in the form of electron transfer and use it to illustrate their ideas [17]. Similar interferences of concepts regarding the structure of ions and ion bonding are experienced in Germany too [5].

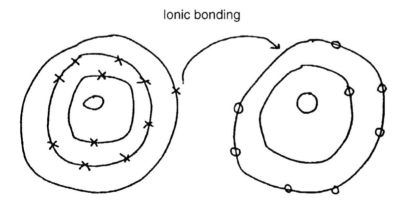

Fig. 5.16 Student's mental model related to ionic bonding [16]

Ion *Formation* and Ionic *Bonding*. The problem in differentiating between the formation of ions (the formation of ions from atoms by redox reaction) and ionic bonding has already been mentioned: students tend to treat the formation of ions via electron transfer and ionic bonding as equal. This can be explained by the fact that the formation of salts is mostly treated in the context of the core-shell model and a redox reaction: therefore, ionic *bonding* is often associated with and treated in the same way as ion *formation*. Naturally enough, the students do not manage to properly separate both concepts. Hilbing [13] determined that "the students connect the concept of the structure of common salt with the formation of sodium chloride from the elements sodium and chlorine. Electron transfer and therefore ion formation from atoms to ions is always a deciding factor in students' images according to the structure of salts".

Based on his observations in class, Hilbing [5] further argues that the explanation of ionic charge through mental subtraction and addition of electrons from atoms tends to place more emphasis on ion formation than on ionic bonding. The subject matter that was taught "belongs completely to the theme of 'ion formation' – i.e. to redox reactions. The ionic bonding and therefore the structure of salts has absolutely nothing to do with redox reactions! In order to describe the model of ionic bonding, the genesis of the charge plays no significant role. (...) The explicit treatment of redox reactions leads to an exaggerated emphasis on terms from the core-shell atomic model: electron, ion formation through electron transfer, usage of the octet rule, electron transfer and ionic bonding through binding electrons. It does not lead to an appropriate idea regarding ions and the structure of salts – one could say that this approach leads up the 'wrong track'" [5].

Also Taber [17] notes that, for students, the ion formation is always combined with the very plausible octet rule. He remarks that students tend to strongly generalize the octet rule and see it as a method with which one can identify stable atoms or ions. In addition, this rule is often used as a general explanation of why chemical reactions take place. He makes it clear that many of the comments and diagrams in schoolbooks encourage the students to look at chemical reactions going on between unlinked atoms in such a way, that they would actively strive towards filling their outer shell "with the help of chemical bonding". In other words, they acquire the shape in which the outer shell is, at each time, filled with eight electrons [16].

Barker [18] additionally comments, that teachers put far too much emphasis on the Octet rule in order to determine formulas and bindings of chemical species. As a result, the students rely on this rule to deduce formulae. During the lesson unit on ionic bonding, teachers often use this rule, in order to show that some atoms fill their shells through electron transfer instead of sharing electrons in covalent bonding. She further points out, that students are not capable of understanding how ion lattices are formed solely based on this explanation [18].

As an addition to Barker's observations, it is determined, that in the lessons, the ion formation is often explained using an example of isolated *single* atoms that are transformed into *single* ions: in this way, neither the spatial structure of

5.3 Mental Models on Ionic Bonding

the original matter nor the structure of the resulting salt crystals is clear. Ionic bonding can only be understood when the spatial structure of salt crystals from *many* ions is demonstrated by models related to the chemical structure of salts: sphere packings or crystal lattices. This connection forms the basis for improvements in chemistry lessons concerning ionic bonding.

Ionic Bonding and Electric Conductivity. The term ion is – as reported – mostly introduced with the help of a redox reaction and ion formation by electron transfer, but also based on experimental observations of electric conductivity or the electrolysis of salt solutions. Hilbing [13] observed such lessons and verifies that these types of measurement are unsuitable for the introduction of the ion: "It seems absolutely appropriate for students of grade 8, with their knowledge of physics, to describe conductivity with the mental model of electrons in aqueous salt solution, as they have learned about the conductivity of metals. They could never come up with the idea of ions" [15] – and certainly not with the idea of ionic bonding!

Sumfleth, Ploschke and Geisler [19] describe that a teacher in his lesson unit introduced ions experimentally through electrical conductivity of various acid solutions. During class, he pointed out several times that the existence of ions in solutions through electric conductivity is proven, while students believe that ionic bonding and therefore electron transfer is substantiated through electric conductivity, that there is no difference in ions and ionic bonding (see Table 5.1).

The conversation shows that the teacher does not clarify the student's position and the misconception is thereby produced or even cemented by the lecture. The student thinks therefore that the electric conductivity of a solution confirms an ionic bonding, whereas for the teacher, it confirms the existence of free moveable ions in the solution. This leads to the student not being able to differentiate between ions and ionic bonding, that students think the conductivity of a solution is equal to that of an ionic bonding: they are not capable of separating ionic bonding from the existence of free moving ions.

Isolated Bonding Between Individual Ions. Taber [15] introduces a further problem regarding the appropriate concept of ionic bonding. In one of his studies, over 100 advanced level students took part and gave their viewpoints on chemical bonding. The result was astonishing when one considers that the students had previously acquired the basics of chemical bonding in several lessons: 60% of the students were of the opinion that a sodium atom can only

Table 5.1 Excerpt from a conversation in a lesson on "Ionic Bonding" [19]

T.	What have we proven?
S.	that the solution conducts electricity...
S.	...an ionic bonding...
T.	So, that's it: there have to be ions in the solution. What could we say about the bonding between the chlorine atom and the hydrogen atom in an HCl-molecule?
S.	Well, I would say that is an ionic bond...
S.	yes, that's true; we have checked that in the previous lesson.

enter into *a single* ionic bond, because it can release only *one* electron from its outer shell. Equally, 58% agreed, that a chlorine atom could also only enter into one ionic bond, because it can only receive one electron into its outer shell [15].

Likewise, Taber reports about students who were given two-dimensional model drawings (see Fig. 5.17) and who were questioned about the number of chloride ions that were connected to one sodium ion. He reports about one of the female students' answers. She stated, at first, that only one chloride ion and one sodium ion could be bonded. After careful consideration of the drawing, she noted however, that each Na^+ ion appears to be surrounded by four Cl^- ions. This confused her because, in her opinion, each sodium particle could only have one ionic bond with one chlorine particle. She was, however, not able to identify one single bond, because she could not say which sodium particle could release an electron to which chlorine particle. Taber therefore assumed that she was of the opinion that the drawing is incorrect. Upon further deliberations, she firmly explained that she could only imagine one single bond between a single chlorine particle and a sodium particle [16].

It is further indicated that misconceptions are mainly because the terms electron transfer and ionic bonding are set equal in chemistry lessons. Students concern themselves only with isolated ions instead of with giant structures. They substantiate the number of bonds that an ion can form, with isolated bonds between individual ions: they reinforce their ideas additionally with the Octet rule and the resulting noble gas configuration of individual ions.

Ion Pairs and Molecules. Students show basic misconceptions regarding ionic bonding even if they are confronted with a correct two-dimensional model of an ionic lattice (see Fig. 5.17) – because they are mostly thinking of ion pairs [15]: one positive and one negative ion are connected through ionic

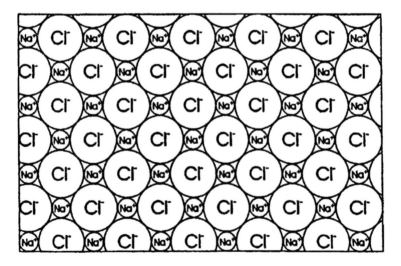

Fig. 5.17 Fitting mental model of the sodium chloride structure [16]

5.3 Mental Models on Ionic Bonding

Fig. 5.18 Misconceptions of ion pairs in the sodium chloride structure [16]

bonding, only weak attraction exists to the other ions in the ionic lattice. Therefore, it is understandable that the students identify individual ion pairs or molecules and only like to visualize exactly two ions bonded together (see Fig. 5.18).

These misconceptions regarding two different types of bonding in an ionic lattice are not only to be found by Hilbing [13] and Taber [15]; the Australian scientists Butts and Smith [20] also identified them in their studies. In addition, they found misconceptions, which, in analogy to scientific historical development, assume that molecules first exist in solid salt crystals, and then the ions develop because of the solution of salts in water.

Consequently, students draw a mixture of ion symbols and molecule symbols in models for an evaporating solution, and use only molecule symbols to illustrate a crystal (see Fig. 5.7). According to the students, "sodium chloride molecules" are formed through the bonding of individual ions with each other and thereby canceling each other's charge: students claim that the ions *neutralize* each other [7]. Another study on chemical bonding by Boo from Singapore [21] led to the fact that students, when discussing the formation of salts, use the argument of neutral particles. Hilbing additionally discovered in Germany, that students postulate electron transfer between the atoms or ions as bond-forming elements and therefore like to speak of "electrostatic neutralized ion pairs" [13].

Based on the above-mentioned students' misconceptions that the ions always arrange in pairs or molecules, it seems understandable that students also have huge problems conceiving the arrangements of ions. The works of Butts and Smith [20] show great deficits of students in spatial images of ionic bonds. They note that most students have a concept of the two-dimensional formation of crystal lattices: "Only four students demonstrated a clear and accurate understanding of the notion of a three dimensional lattice of sodium ions and chloride ions, although more were able to draw two dimensional diagrams which indicated that they did have only some understanding of the crystal structure of salt" [20]. Hilbing also came to similar findings: although several of his examined students have concepts relating to infinite, two-dimensional arrangements of particles, most students draw a small number of ions thereby reflecting finite molecules [13].

Animistic Concepts of Ions. From the literature, the animistic idea of ions, i.e. their personification, has been named as animistic misconception. This can be equated with the anthropomorphic language with which the "wishes or preferences of ions" are described. As an example, students state that electrons *belong* to certain atoms, or they formulate their statements regarding two atoms in such a way that an atom *wants* to form a bond, or an atom *would like* to receive an electron because it *wishes* to have a full outer shell. Taber continues with "atoms – according to students – *like* to be stable, *wish* to be stable, *prefer* to be stable and indeed can be *very eager* to be stable" [17]. Various scientists in relation to both ionic formation and ionic bonding [16, 22, 23], have also found such an animistic language used by students.

However, this phenomenon does not really need to be followed-up. It does not constitute a misconception in the classic sense in that it does not hinder the lesson or development of scientific understanding in any large way. On the one hand, many teachers also use this animistic description of phenomena, on the other hand many young people are aware of the fact that ions are not living creatures and use the personification merely in order to graphically describe such terms. The animistic form of expression can be useful for the students as a bridge to an appropriate subject terminology, because it allows them to describe the phenomena intuitively and with their own familiar language. In this way, they lose their inhibitions and are able to describe unknown phenomena at a very simple level. If the teacher wants to improve their terminology he should rather use the correct one consistently: this may replace such personified language and may guarantee better learning.

Teaching and Learning Suggestions. In the historical development concerning the concept of the structure of matter, the atom was first postulated (John Dalton, 1808) and much later the ion (Svante Arrhenius, 1884). The same applies to the concept of "atomic bonding" and "ionic bonding". Using his valence theory, Kekulé in 1865, started describing many molecules in organic chemistry in such an appropriate way that these descriptions are still apt today. If the historic development of these mental models is to be thematically dealt with in today's lessons, then, they should be kept in chronological order. If one chooses not to argue from a historical point of view, but rather from a chemistry education view point, the chronological order remains open. Reiners [24] recommends that one should first teach covalent bonding and then should deal with ionic bonding as a borderline case. She does not follow Taber [25], but quotes from the well-known textbook in Inorganic chemistry of Huheey [26]. A reader's letter answered her with arguments from Huheey himself: "'Because pure ionic bonding can be described

5.3 Mental Models on Ionic Bonding

using the electrostatic model, it is advantageous to treat it first'. Huheey makes it clear so as to avoid learning difficulties in the lessons concerning chemical bonding: 'Ionic bonding can already be introduced at the Dalton model level, taking the ion into consideration and the electrostatic attraction or repulsion'. The core-shell model is of no help whatsoever for the ionic *bonding*, i.e. the connection of ions. For understanding, the core-shell model is indispensable in the *formation* of ions from atoms and vice versa" [27]. Huheey's argument in which he recommends teaching ionic bonding without the core-shell model is thereby strongly supported.

However, Reiners opposes this and states, that in this way "the formation of ions cannot be correctly explained" [28]. This argument leads to the following statement from Sauermann [29]: "Behind this sentence lies a type of explanation enforcement, with which many young chemistry teachers have been inoculated at universities. It results from the enforced concept that ions have to be formed from atoms, which is necessary before the formation of the atom can be 'explained'. I can only describe this procedure as a sham package: take the students to a mine, observe salt crystals, describe them with existing ions in their force quality to form the well-known ionic lattice".

As we have already mentioned (see Sect. 5.2), it is recommended, to set "atoms and ions as a basic particles of matter" based on Dalton's atomic concept. We can also introduce ionic bonding with the special Periodic System (PSE). Combining ions, "ions left and right in the PSE" (see Fig. 5.10) ion lattices and ionic bonding are introduced. Either the first combinations of ions could be looked at with the help of model drawings (see Fig. 5.11) and the attraction of differently charged ions or the repulsion of similarly charged ions could be compared and verbally described. Alternatively, one could produce closest sphere packings using two different types of spheres (see Fig. 5.13) in order to show the real structure of the sodium chloride crystal: the fixture of the ions in the crystal could be elucidated using non-directed electric energies surrounding each ion.

Should one attempt to demonstrate the non-directed electric forces of the ions, many experiments could show electrostatic attraction or repulsion. One example is to rub a new overhead transparency on a normal piece of writing paper thereby observing the attraction, or to rub two transparencies and demonstrate repulsion (see E5.13). It is also possible to take small round magnets of two different sizes. Using one with the north pole facing upward and the other with the north pole facing downward, place them in a flat see-through box and shake on the overhead projector (see E5.14): the magnets form a pattern, which can be compared to a cut-out layer from the sodium chloride structure. The non-directed forces surrounding each of the magnets cause a close packing of the disks with the coordination number 4 (see Fig. 5.19). It should however be noted that magnetic forces are at work in this model, the ions in a salt crystal are however held together through electric forces.

If one finally wishes to demonstrate the non-directed force field, which surrounds a magnet, one has the possibility of putting iron filings on a glass plate and placing a magnet under it (see E5.15). When the glass plate is briefly

Fig. 5.19 Model experiment for non-directed forces using magnets of two types

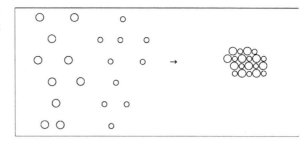

tapped, the iron filings form a specific pattern and show that magnet forces go into all directions (see Fig. 5.20). Once again, the demonstrated effects are of a magnetic kind; in salt crystals, there are non-directed electric forces holding ions together.

In order to expand on and demonstrate the utilization of ionic bonding one can show experiments on growing crystals (see E4.1), the solubility of salts, on spontaneous crystallization from saturated solutions (see E5.11) or precipitation reactions (see E5.12). In all cases, rearrangements of ions take place. If one not only formulates abstract reactions in this topic but produces two-dimensional model drawings with related ions (see Figs. 5.11 and 5.12), then the students have additional visual learning aids concerning ionic bonding.

The above-mentioned reactions have been chosen such that the involved ions are retained and only their arrangement is changed: at this stage, no traditional redox reactions are to be demonstrated and interpreted. In this way one can teach ionic bonding, but avoid the undesirable connection of the terms ion *bonding* and ion *formation* – and ease the learning. Additionally ions are introduced at an early stage as an important particle type alongside atoms and molecules: "One does not describe the structure of salt crystals from 'imagined molecules' or 'connecting units' or 'modified atoms' or 'connectivity particles' thereby circumventing the term ion" [30]. On the contrary, one looks at atoms and ions as essential "basic particles of matter" which constitute all substances on earth!

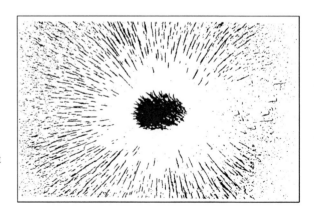

Fig. 5.20 Model experiment for non-directed forces surrounding a magnet (see E5.15)

5.4 Chemical Structures and Symbolic Language

"H_2O freezes at 0 degrees and has a bent angle; Na melts easily and possesses 11 protons; an atom has a core, the electrons fly around it".

If one has differentiated information about elementary particles like protons, neutrons and electrons which connect to questions of chemical bonding, misconceptions can arise which mix bent water molecules or the 11-protons nucleus of a sodium atom with macroscopic characteristics of matter (see also Sect. 4.3). Since electrons are not ordinary basic particles of matter in the sense of atoms, ions and molecules, but are recognized more as charged clouds, orbital or through the particle-wave duality, the mixing of macroscopic and sub-microscopic characteristic properties should be avoided more carefully.

This mixing of such levels of property occurs consciously or sub-consciously even amongst teachers and professors. For example, an organic chemist might do the following as he or she explains the mechanism of an electrophilic substitution reaction, the bromination of benzene: "the bromine approaches the benzene (...) bromine attacks the benzene core (...) the electrons relocate and the bromine splits (...) bromobenzene results (...) a bromine has substituted a hydrogen".

The lecturer *means*, of course, an attacking Br_2 molecule; he or she *means*, of course, the formation of a monobromobenzene molecule; and finally, he or she *means*, of course, that H atoms are substituted by Br atoms. Everyone knows what he or she *means*: the lecturer mixes up matter and particles, and resorts to the well-known "laboratory jargon". This laboratory jargon is understood, of course, amongst laboratory assistants and chemists. However, it cannot be of much help to learners who are introduced to special terminology for the first time. One should make an effort to say what one means, especially with beginners; one should distinguish between Br_2 molecules and Br atoms, and in speaking of H atoms or Br atoms in bromobenzene molecules – if that is what one means!

The way in which one speaks of electrons should be left completely open. If one is assessing the number of electrons of an atom or molecule, then electrons are considered more as elementary particles. If however, electrons are considered in chemical bonding, one should rather discuss the electron or charge cloud, may discuss centers of electrical charge and dipolar molecules.

There are ample self-made misconceptions regarding the formulation of reaction symbols. Mulford and Robinson [32] discovered the following situations regarding questions 5 and 6 (see Fig. 5.21) when evaluating the empirical studies: "Responses to question 5 suggest that students came to us with a very poor understanding of chemical formulas and equations. Only 11% selected the correct answer d. When we consider the number of students who selected responses a, c and e, we see that 65% chose responses that do not conserve atoms. Combining responses a, b and e indicates that 74% appear not to understand the difference between the coefficient "2" and the subscript "3" in 2 SO_3" [32].

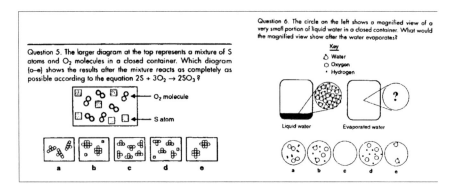

Fig. 5.21 Two questions on chemical symbols and mental models [32]

In "Question 6" of the study, students were asked, which particles are present after water is boiled (see Fig. 5.21): "Only 39% chose widely separated molecules, 37% selected response a, b or d, which represent, respectively, H_2 and O_2 molecules, a mixture of H_2O molecules and H atoms, and O atoms, respectively. Twenty-four percent chose response c, with no particles at all" [32]. The well-known H_2O formula continually tempts young people to think of H_2 and O_2 molecules when observing boiling water – even when they know the oxygen-hydrogen reaction!

In relation to the formula for ionic lattices, Hilbing [13] and Taber [34] clearly recommend considering the importance of the formula indices with the students and interpreting the ion number ratio accordingly. The formulas describe the number ratio of ions, which enter into electrically balanced ionic lattices – for example, Ca^{2+} ions arrange with the double number of Cl^- ions: the formula $(Ca^{2+})_1(Cl^-)_2$ results (see also Fig. 5.11).

In a chemical equation, the numbers in front of the separate atoms, ions or molecules are known as coefficients: they indicate exactly how many of the atoms, ions or molecules react with another type of atoms, ions or molecules in order for a complete reaction. As an example, let us look at the well-known reaction of calcium chloride solution with silver nitrate solution to form insoluble silver chloride:

$$1\,CaCl_2(aq) + 2\,AgNO_3(aq) \rightarrow 2\,AgCl(s) + 1\,Ca(NO_3)_2(aq)$$

In this reaction, essentially only silver ions and chloride ions join to create an ion lattice:

$$2\,Ag^+(aq) + 2\,Cl^-(aq) \rightarrow (Ag^+)_2(Cl^-)_2(s)$$

$$or \quad Ag^+(aq) + Cl^-(aq) \rightarrow Ag^+Cl^-(s)$$

5.4 Chemical Structures and Symbolic Language

Schmidt [33] finds further misconceptions in his empirical studies related to the assignment by the student of the various formulas. On the one hand, he mentions that students do not have an appropriate concept of ions, which creates problems for them. Students are often not able to write down the involved ions in a given ionic compound. As an example, it is found that students, studying Na_3PO_4, imagine the substance as being broken down into Na^+, Na_3^+, Na_3^{3+}, PO_4^{3-}, PO_4^-, $NaPO_4^{2-}$, or $Na_2PO_4^-$ as well as completely in the individual elemental ions: Na^+, P^{5+} and O^{2-} [33]. Many students are consequently not able to appropriately break down ionic compounds in the involved basic units of matter.

On the other hand, errors emerge regarding the mass ratio for deriving formulas, i.e. the ion number ratio. Schmidt [33] gives us the following example according to the ratio of sodium ions to carbonate ions in sodium carbonate solutions. Students tend to inappropriately set the mass ratio $m(Na): m(CO_3)$ equal to the number ratio $n(Na^+ ions) : n(CO_3^{2-} ions)$ in the compound.

Accordingly, students often round off the mass ratio 46:60 generously to a quantity number ratio of ions to 2:3, which has a very different meaning. Advanced high school students are often not able to derive at the actual ion number ratio of 2:1 of sodium ions and carbonate ions from the mass ratio.

In addition, it seems as if students tend to confuse formula indices, charges and coefficients massively and thereby end up using them inappropriately. Schmidt quotes several examples of specific misconceptions that students show in the interpretation of the formula M_2CO_3 where M means one atom of an alkali metal. They don't know the basic units of the ionic compound; they interpret the formula as $M_2(CO)_3$ and derive an ion number ratio of 2:3. As a result, they set two different ratios equal that are not consistent with each other at all. They misplace the ratio of the formula indices $a:b = 2:3$ as the mass ratio [33]. In this respect, Hilbing [13] summarizes some difficulties of the difference between the coefficients and formula indices (see Table 5.2).

Teaching and Learning Suggestions. The problem of the formula indices and factors is a constant one in lessons on chemical terminology or related to specific chemical symbols. This problem goes so far that learners, who do not understand the formula language, end up developing a negative attitude to chemistry or to lessons in chemistry. Strehle [35] was able to prove significant correlations by German students in grades 9 and 10, thereby confirming the hypothesis: "There are correlations between the attitudes of students and

Table 5.2 Excerpt from a lesson conversation on "Factor and Index" [13]

T:	We could now combine a metal with an arbitrary halogen. Who can name that salt if I am going to take magnesium and a halogen?
Paul:	MgCl…
T:	try it again…
Paul:	…two MgCl…
T:	It doesn't really matter to me if you even take 15. But MgCl appears to be incorrect: look at the Periodic Table of the elements.
Paul:	MgCl two.

understanding formulas of chemistry lessons" [35]. Not only does a good understanding of the symbolic language lead to a better understanding of chemistry, but also a positive attitude toward chemistry can be developed – it is really an intrinsic premise for success in chemical education!

As is shown in the last example of the previous chapter, problems with the symbolic language get worse when ionic symbols and the balance of protons in atom cores and electrons in the shell are treated. Electron transition, the Octet rule and noble gas configuration cause more confusion than clarity in this respect. Difficulties can be avoided introducing the ion in the sense of the Daltonic model (see Sect. 5.3) and ionic symbols in connection with the special Periodic Table (see Fig. 5.10).

On the other hand, lecturers or teachers for beginners should carefully separate the language levels in their lessons – for example (see also Sect. 2.5):

– Please speak about ice, water and steam at the *macro level* – not from H_2O!
– Speak of water particles or H_2O molecules at the *sub-microscopic level*,
– speak on this level of white and red spheres as models for H atoms and O atoms,
– speak of a model for one H_2O molecule – and not of the model for H_2O or even water!

With respect to the misconceptions in this sub-chapter, water can be described as having a freezing point of $0°C$ or the density of 1 g/ml, the "angle" does not relate to the characteristics of the substance, but rather of attributes of the H_2O molecule and should be associated with the language of the particle level: H_2O molecules are constructed angular.

Sodium has a melting point of $98°C$. The mentioned 11 protons are not to be related to "sodium" but to the nucleus of one Na atom or one Na^+ ion. It is important to carefully differentiate these different language levels at the beginning of lessons. Teachers should note the differences carefully so that students avoid the potential misconceptions.

Other useful aids in setting up appropriate conceptions regarding ion terminology are visualizations, for instance, in the form of model drawings (see Figs. 5.11 and 5.12). If additional ion symbols of salts in real number ratios are added to the sketches, then students who are not very good at abstract thinking would have a chance of setting up appropriate conceptions. Regarding the difficulties with formulas of salts (see Table 5.2) one could sketch, for example, ionic symbols in the number ratio for *solid* salts in a regular, for the *solution* in an irregular pattern (see Fig. 5.22). If, for instance, mineral water [8] is treated in lessons and the related smallest particles are discussed, then it is important to present a model drawing of a stylized beaker and to sketch in the related symbols for molecules like H_2O and CO_2, respectively for ions like Ca^{2+}(aq), Mg^{2+}(aq), Na^+(aq), Cl^-(aq), SO_4^{2-}(aq), HCO_3^-(aq) (see Fig. 5.22). Through further model drawings for drinking water and seawater one could visually demonstrate the associated differences in composition.

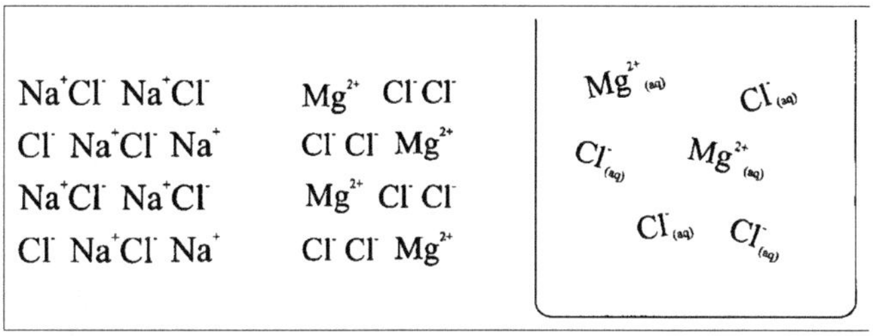

Na+Cl- Na+Cl-
Cl- Na+Cl- Na+
Na+Cl- Na+Cl-
Cl- Na+Cl- Na+
Mg2+ Cl-Cl-
Cl- Cl- Mg2+
Mg2+ Cl- Cl-
Cl- Cl- Mg2+
Mg2+(aq)
Cl-(aq)
Cl-(aq)
Mg2+(aq)
Cl-(aq)
Cl-(aq)

Fig. 5.22 Ionic symbols for visualizing structures of salts and salt solutions

If one wants to supplement the formal drawings for the two dimensional models, sphere packing models for the sodium chloride structure as well as spatial lattice models are suitable (see Fig. 5.13). There are also red-green 3-D pictures available that appear three-dimensional when looked at through the red-green glasses [34]. The students are mostly very motivated in using these 3-D pictures and enjoy playing around with them for a while. If one also marks the related ions with their symbols in the structures of these 3-D pictures of salt crystals and the students transfer these sketches into their notebooks, they will tend to remember ions as basic particles of salts. Again, this kind of visualization will certainly help them understand the whole process better.

Likewise, with respect to the example of silver chloride or barium sulfate precipitation from respective solutions, ionic symbols in model drawings are more appropriate than formal reaction equations. One could first name and sketch the ions of the starting solutions *ahead* of the precipitation reaction, and could mark them with (aq) symbols. Only after carrying out the precipitation could one consider the type of solid matter and set the symbols for the related solid matter and the remaining solution (see Fig. 5.23).

There are quite helpful similar models to show the solutions before and after the neutralization. From experience, misconceptions regarding "HCl molecules" in hydrochloric acid and "NaOH molecules" in sodium hydroxide solution tend to play much too large a part. This is why one should avoid speaking solely of HCl as an "abbreviation" for hydrochloric acid and of NaOH as an "abbreviation" for sodium hydroxide – when possible one should mark the ions in the solutions. Even more important are model drawings that, in stylized beaker models, state the type of ions using (aq)-symbols (see Fig. 5.24). Of course, it is necessary

Ba2+(aq) Cl-(aq)
Cl-(aq)
Cl-(aq) Ba2+(aq)
Cl-(aq)
+
SO4 2-(aq) Na+(aq)
Na+(aq)
Na SO4 2-(aq) Na+(aq)
→
Cl-(aq)
Na+(aq)
Cl-(aq) Na+(aq)
Cl-(aq)
Na+(aq)
Ba2+ SO4 2-
SO4 2- Ba2+
Cl-(aq) Na+(aq)

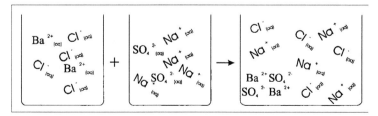

Fig. 5.23 Ionic symbols for visualizing a precipitation reaction (see Fig. 5.12)

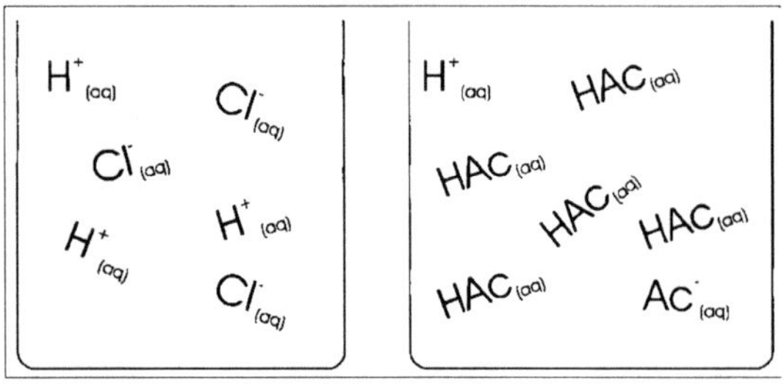

Fig. 5.24 Model drawings for "strong hydrochloric acid" and "weak acetic acid"

$H^+_{(aq)}$ $Cl^-_{(aq)}$
$Cl^-_{(aq)}$
$H^+_{(aq)}$ $H^+_{(aq)}$
       $Cl^-_{(aq)}$

$H^+_{(aq)}$ $HAC_{(aq)}$
$HAC_{(aq)}$
         $HAC_{(aq)}$ $HAC_{(aq)}$
$HAC_{(aq)}$ $AC^-_{(aq)}$

to on the one hand discuss the drawings of beakers as models, and on the other hand, to reflect the ionic symbols as models for moving ions in acidic, basic and salt solutions (see also Chap. 7).

Model drawings, which show the difference between strong and weak acids, are very important (see Fig. 5.24). One should use only ionic symbols to demonstrate a strong acid, and the mixture of ionic symbols and molecular symbols to visualize a weak acid. To neutralize weak acetic acid with sodium hydroxide solution, the reaction of the HAc molecules with OH^-(aq) ions, *and* the reaction of H^+(aq) ions with OH^-(aq) ions, should be compared and visualized accordingly (see also Chap. 7).

Finally, reactions of the salts, chemical structures and ionic symbols could be simulated and animated on the computer by using certain computer programs. There are interactive learning programs for these themes that are particularly motivating for young learners in chemistry. Moeller [35] created the learning environment, "Structures of Metals and Salts", that interactively introduces terms like ionic lattice, elementary cube and unit cell. The users cannot only look at sphere packing, crystal lattice and unit cell, but they can playfully manipulate them, thereby discovering structures and formulas derived from structures [36].

Perhaps the virtual computer programs could be more easily considered as models in a scientific sense for students and do not exhibit the irrelevant items like the models using spheres, rods and glue. We can anticipate a selection of these newly developed computer programmes and learning exercises – a good selection of *many* examples for *one and the same* theme would be a great advantage in understanding chemical models!

5.5 Experiments on Structure–Property Relationships

E5.1 Silver Crystals Through Electrolysis

Problem: Metals cannot easily be visibly crystallized because they form very small crystallites and thus arbitrary conglomerates. Sometimes, zinc crystals can be found on polished zinc posts that look like "ice flowers" on a windowpane. One often finds beautiful rainbow-colored crystallized cubes of crystal cascades of molten bismuth in mineral collections. One possibility of attaining metal crystals in the school laboratory can be achieved by cementation (E4.4). Another possibility is growing silver crystals by specific electrolysis of a silver salt solution – this can be well observed on the overhead projector.

Material: Glass bowl, alligator clips, cable cord, iron wire as electrodes (take combustion spoons), 10–20 V source (or serially connected batteries); dilute silver nitrate solution, concentrated ammonia solution.

Procedure: Add few milliliters of ammonia solution to approximately 100 ml silver nitrate solution until the initially formed precipitate dissolves. Place the solution in a glass bowl. Fix two iron wires in such a way that one wire is in the form of a round electrode; dip the other one, the negative electrode, at the middle of the bowl. Apply a voltage of 10–20 V DC to the electrodes.

Observation: Cascades of crystals form from the central electrode: silver crystals (see also Fig. 5.2).

E5.2 Closest Sphere Packings as Models for Metal Crystals

Problem: The structure of many metal crystals can best be described by using closest sphere packing models. Because there are two possibilities of systematically sphere packings, these should be introduced first (see also Figs. 5.3 and 5.4). Both packings can be differentiated by the layering sequence ABAB and ABCA – with respect to the densely packed triangular layers.

Material: Cellulose spheres (30 mm), glue.

Procedure and Observation: (a) First, build triangles made of 3 spheres and hexagonal layers made of 7 spheres (see picture). Layer them on top of each other in two different layering sequences: ABAB relates to the hexagonal closest packing, ABCA relates to the cubic closest packing (see Fig. 5.3). Determine the coordination number for each: one sphere is touched in each case by 12 other

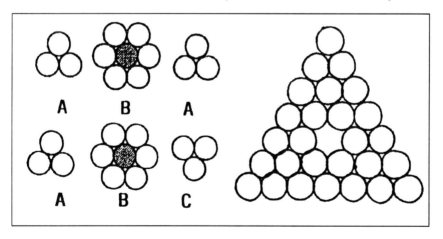

spheres. **(b)** Make an equilateral triangle with a side-length of 7 spheres with a hole in the middle (see picture). After that, layer spheres using the sequences ABAB and ABCA: two different types of spherical pyramids result (see Fig. 5.4).

Tip: Sphere packings with the layering sequence ABAB of triangular layers are known as the *hexagonal* closest packing; this is found in the metals magnesium and zinc. If sequence ABCA of triangular layers is realized, this shows the *cubic* closest packing, which is present in copper, gold, silver, lead, nickel and many other metals.

E5.3 Cubic Closest Packing as a Model for Silver Crystals

Problem: Of the two possible types of closest packings, the *cubic* closest packing is the appropriate model for a silver crystal: Ag atoms are arranged in silver crystals like spheres in the cubic closest packing. The packing is called cubic because it contains a special packing called elementary cube. It should be constructed and integrated in the cubic sphere packing model. The unit cell is constructed using the elementary cube as a basis – as a preparation for cubic cells for salt crystals (see E5.8).

Material: Cellulose spheres (30 mm), glue, Gummy candy, cardboard (see picture).

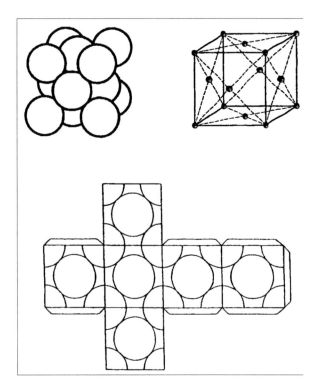

Procedure and Observation: (a) Make an equilateral triangle with a side length of 7 spheres with a hole left in the middle (see picture E5.2), layer as many spheres as possible on top of the triangle using the sequence ABCA: closest sphere packing in tetrahedral form (see E5.2 and Fig. 5.4). **(b)** Glue together an **elementary cube** out of 14 spheres, either "1 + 6 + 6 + 1" or "5 + 4 + 5" spheres (see picture and Fig. 5.5). **(c)** Place the cube into the sphere packing (a): it is standing on a sphere of one corner of the cube (filling the hole), the cube is directed up along one spatial diagonal, the sequence order is ABCA related to the packed trigonal layers. Add other spheres to show the *cubic* **closest sphere packing** (ABCA). **(d)** Pack spheres of same size densely in square layers; build the elementary cube into the close packing: the cube is standing on one face (see Fig. 4.4). **(e)** Construct the **crystal lattice** of the elementary cube using Gummy candy and toothpicks (see picture). After a few days the candy becomes very hard, which allows the model to be used a long time. **(f)** Enlarge the cube net (see picture) on handicraft cardboard, exactly cut out and glue together to form a model of the **unit cell** (see Fig. 5.5): all sphere parts of the unit cell together make up 4 complete spheres.

E5.4 Ductility of Different Metals

Problem: Structure–property relationships can be shown through comparing the ductility of metals of the cubic face-centered structure, and of the hexagonal structure. In the crystallites of copper and gold, there exist close packed atom layers of the cubic closest structure. Starting from the elementary cube, the close packed triangular layers can be moved vertically in four different directions related to the four spatial diagonals of the cube. When a force from arbitrary direction sets in, triangular layers exist and can be moved: gold leaf or copper leaf can be seen. In crystallites of magnesium or zinc, there is only one preferential direction for moving close packed triangular layers. Due to the hexagonal structure of magnesium and zinc crystals, there is no "magnesium leaf" or "zinc leaf".

Material: Solid steel mats, hammer; copper plate, zinc plate, magnesium ribbon.

Procedure: Hammer all three metal samples in such a way, that the material strength is continuously reduced.

Observation: The surface of the copper piece increases; it is flattened to copper leaf. The magnesium or zinc sample disintegrates in bits after being hammered for some time, it crumbles into little metal chips.

E5.5 Formation and Decomposition of Sodium Amalgam

Problem: Alloys are produced because they have different properties than the original metals. One can produce steel with certain properties according to the individual wishes of the metal usage. As an example, let us look at the formation of sodium amalgam from its elements: sodium is silver-colored and soft. Mercury is also silver-colored and liquid at room temperature. Sodium amalgam is dark grey and very brittle, it breaks at the slightest force. The flash of light in the reaction of sodium and mercury shows an exothermic reaction to take place – an argument against the idea that alloys are not formed in chemical reactions but rather mere mixing processes. Dentists also produce certain alloys from silver, zinc and mercury, in order to obtain a moldable mass which fills the hole of the tooth and becomes quite hard after some time.

Material: 200 ml beaker, test tube, pipette, glass rod, tweezers, knife; sodium, mercury, universal indicator paper.

Procedure: Carefully place a drop of mercury into a test tube using the pipette. Place the test tube in a beaker to prevent any spillage. Freshly cut a pea-sized piece of sodium, place in the test tube and press into the mercury drop using the glass rod. Place the resulting sample of solid matter in the beaker and break with the tweezers. Add several ml of water and test the solution with indicator paper.

Observation: A flash of light can be observed when the sodium is pressed into the mercury, a dull grey solid matter results which is quite brittle and which can be broken: sodium amalgam. In the reaction with water, small bubbles (hydrogen) occur and a solution is produced that colors the indicator paper blue: sodium hydroxide solution. Liquid mercury remains after some time, it can be put back into the container.

E5.6 Nitinol – A Memory Metal

Problem: In 1965 some scientists produced a very special nickel-titanium alloy which can remember its programmed form. They called this kind of alloy "memory metal" and the special compound "Nitinol", because it was discovered in the **N**ickel-**Ti**tanium **N**aval **O**rdonnance **L**aboratory. The memory metals are often used in engines, motors and other pieces of equipment, for which it is important to have a certain form at a certain temperature, e.g. for closing a valve. The conversion temperature of the Nitinol composition Ni_1Ti_1 is approximately 50°C. Samples of such memory metals, like nitinol wires, can be obtained from Educational Innovations (teachersource.com).

Material: Beaker; samples of memory wire, memory metal pre-shaped springs.

Procedure and Observation: Bend a wire into any shape and hold it in the flame of a match or candle: the wire spontaneously regains its shape (demonstrate it by using an overhead projector). Dip the wire which is re-formed into hot water and observe. Either press together or stretch out springs and place into hot water (overhead projector): once again the original form returns.

Tip: A nitinol wire can be reprogrammed to different shapes: the new wire form has to be stabilized between two steel plates and heated red hot with a burner for several minutes. If an oven is available, the steel plates with the fixed wires can also be heated for 10 min at approximately 500°C. This new form always reemerges whenever the wire is reshaped at room temperature and then heated until the transformation temperature is exceeded.

E5.7 Freezing Point Temperatures of Solutions

Problem: Water freezes in normal conditions at exactly 0°C. Dissolved substances lower the freezing temperature of the solution based on the concentration of the solvent: 1-molar glucose solution freezes at −1.9°C, 2-molar glucose solution freezes at −3.8°C. One can state that the higher the concentration of a solution, the lower the freezing temperature. This fact can be used as an experimental introduction to the concept of ions. A one molar table salt solution does not deliver the expected value of −1.9°C, but rather doubles the value of approximately −3.8°C. The analysis shows, that 1 l of 1-molar NaCl solution contains 2 mol particles: 1 mol sodium ions and 1 mol chloride ions. One can state that 1-molar sodium chloride solution is 2 molar with respect to the total number of ions.

Material: Glass bowl, test tubes, thermometer; glucose, sodium chloride, ice.

Procedure: Prepare 1 molar solutions of glucose and sodium chloride, and add several milliliters in two test tubes. Place the solutions in a cold mixture of ice and table salt (−15°C) and stir with a thermometer. For observational purposes, quickly remove the test tubes from the mixture and as soon as the first crystals appear, measure the temperature.

Observation: The glucose solution freezes at approximately −2°C, the sodium chloride solution at approximately −4°C.

Tip: 1-molar calcium chloride solution ($CaCl_2$) could be tested as a further example: it freezes at approximately −6°C. 1 l of the 1-molar solution contains 3 mol of ions, 1 mol calcium ions and 2 mol chloride ions.

E5.8 Structural Models for the Sodium Chloride Structure

Problem: The structures of many metal crystals correspond to the cubic closest packing (see E5.3). This packing contains rather large holes, which are formed by 6 spheres: octahedral interstices (see picture). In addition, smaller gaps that are created by 4 spheres can be examined: tetrahedral interstices (see picture). The structure of many salts can best be described using the closest sphere packing of anions, in which the gaps are partly or completely filled by smaller cations. The structure of sodium chloride crystals can be described as the cubic closest packing of chloride ions in which all octahedral interstices are filled in with sodium ions. The lattice can be shown as a short cut to the close packing model, because it only shows the positions of the ions and no longer their relative size.

Material: Cellulose spheres (white 30 mm and red 12 mm), glue, cardboard, Gummy candy (two different colors), toothpicks.

Procedure and Observation: (a) Add white spheres on the triangular layer in the cubic closest packing (see E5.3), find the shape of both interstitial sites: octahedral and tetrahedral gaps. (b) Completely fill the octahedral sites with the

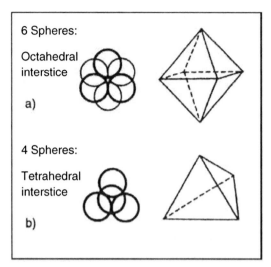

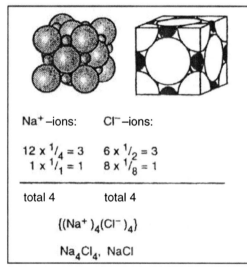

small red spheres: **close-packing model** of the sodium chloride structure (see Fig. 5.13). The coordination number 6 is established for both the large and the small sphere. (**c**) Produce the **elementary cube** (see picture) using 14 white and 13 red spheres and build the cube into the packing (see Fig. 5.13). The elementary cube shows the cubic form of the salt crystals; it also shows the relative size of ions and their positions – it does not show the ratio number of ions 1:1. (**d**) Imagine dividing the outer spheres of the elementary cube vertically through the spheres (see picture) and obtaining a **unit cell** (styrofoam balls can be cut by passing a hot wire through them). It is easier to build the unit cell from a cube net, which has been drawn on cardboard (see E5.3 f). Counting the unit cell leads to the ion number ratio 4:4 or 1:1 (see picture). **(e)** In order to get a better view inside the structure it is also common to use the **crystal lattice** as a structural model (see Fig. 5.13). It only shows the positions of the ions. Construct a lattice using two differently colored Gummy candy and toothpicks (see E5.3 e).

E5.9 Destruction of Salt Crystals

Problem: If one hammers on a sample of metal (lead or copper), the metal can be flattered to lead or copper leaf: metals with the cubic closed structure are very ductile (see E5.4). If, on the other hand, one hammers on a rock salt sample, the crystal splits into tiny pieces or breaks apart, forming crystal plates. These properties can be explained through the structure: the layers of ions in a crystal are moved when force is used; similarly charged ions stand at opposite sides and are responsible for the repulsive effect of the crystal layers (see Fig. 5.14).

Material: Hammer, tile as a trivet, knife; copper or lead sample, rock salt sample.

Procedure: Hit metal samples as well as salt crystal samples with a hammer. Treat the salt crystal with a knife too: place the knife vertically on a piece of rock salt and then hit the knife.

Observation: Metal samples are shaped in different forms based on the direction and force of the blow. Salt crystals break into cubic shapes when hammered, by the knife thin layers of crystal may break off parallel to the cube.

E5.10 Electric Conductivity of Salt Crystals, Melts and Solutions

Problem: Salts are substances, which are composed of ions – because of their charge the ions are capable of moving in an electrical field. But the ions are statically built into solid crystal and cannot move: no conductivity can be ascertained at room temperature. As soon as a big crystal is strongly heated the ions become movable. This can be proven by using conductivity measurements. In melt and solutions, the ions are quite movable: electric conductivity results. Because sodium chloride has the very high melting temperature of 800°C, a mixture of lithium chloride and potassium chloride are molten together and used.

Material: Tripod and wire gauze, tripod and clamps, burner, two iron nails, cable wire, alligator clips, porcelain crucible, 9-V battery, glass beaker, graphite electrodes, multimeter (V and A), lamp; rock salt crystal, lithium chloride, potassium chloride.

Procedure: **(a)** Firmly secure a rock salt crystal in between two iron nails on a wire gauze; connect both nails to the 220 V or 110 V plug and with the multimeter (caution!). Strongly heat the crystal while observing the multimeter. **(b)** Make ready an electrical circuit of battery, multimeter, lamp and two graphite electrodes. Intensely heat a mixture of 21 g lithium chloride and 7 g potassium chloride in a crucible until a clear molten liquid is attained. Dip both graphite electrodes into the molten liquid. **(c)** Repeat the test again using solutions of all three salts.

Observation: **(a)** The crystal does not conduct electricity at room temperature, but does so when strongly heated. Where the crystal touches the iron nails, one sees light flashes and color changes in the crystal. **(b)** Molten salts conduct electricity even at a voltage of 6 V: the lamp light up, the multimeter shows certain electrical conductivity depending on the distance between the electrodes from each other. **(c)** Solutions of salt conduct electricity.

E5.11 Heat of Crystallization from Molten Salt

Problem: When crystals melt certain energy is required in order to separate ions from an ion lattice. The same energy is set free when ions from the molten liquid are added to form solid salt crystals. This can be convincingly demonstrated when a cooled-off molten salt, i.e. sodium acetate hydrate, remains liquid even at room temperature. It only crystallizes again, when a seed crystal is added. For this reason, these salts are used in little heat pillows which produce a pleasant warmth by pressing a button; they are also known as "pocket warmers". Resulting solid crystals can be transformed back into the cool molten matter by heating first in boiling water and then cooling down to room temperature.

Material: Small flask, burner, thermometer, tripod and clamps; sodium acetate hydrate.

Procedure: Half-fill a flask with sodium acetate hydrate and melt the salt completely through the use of the burner. Leave the molten matter to cool in the air, insert a thermometer and measure the temperature. Add one crystal of sodium acetate hydrate to the liquid, observe first, and measure the temperature again.

Observation: The molten matter crystallizes instantly when the seeding crystal is added. The temperature of the crystalline mass rises during this procedure to more than 50°C. The test can be repeated through continued melting and cooling down.

E5.12 Precipitation of Salt Crystals from Solutions

Problem: In demonstrating the precipitation of crystals, it is possible to mix certain solutions: the combination of two particular types of ions lead to nearly insoluble salt crystals (see for example Fig. 5.23). In an initial interpretation, one would speak of insoluble salts, they are however usually soluble in very small quantities and are therefore known as slightly soluble – the solubility product will quantitatively provide the limited solubility. Since certain deposits

5.5 Experiments on Structure–Property Relationships

are easily recognizable, one speaks of detection reactions for certain ion types: chloride detection by precipitation of silver chloride, or sulfate detection by precipitating barium sulfate.

Material: Test tubes; diluted solutions of silver nitrate, sodium chloride, hydrochloric acid, barium chloride, sodium sulfate, sulfuric acid.

Procedure: (a) Mix small volumes of diluted silver nitrate solution with several drops of other solutions listed above. (b) Mix small volumes of barium chloride solution with a few drops of the other solutions listed above.

Observation: (a) A cheesy-white deposit results from silver nitrate solution when hydrochloric acid or chloride solutions are added: chloride detection. (b) A crystalline white deposit is left from barium chloride solution when sulfuric acid or sulfate solutions are added: sulfate detection.

E5.13 Electric Attraction and Repulsion Forces

Problem: It is not possible to see gravitational, magnetic or electric fields – this is one of the basic difficulties in judging attraction forces. It is, however, possible to sense them: earth's gravity forces all objects to fall to the ground, magnets strongly attract iron objects. One speaks of electric forces when one has been electrically charged by a carpet or when one sets a slight shock on a door handle. These electric forces work as connecting forces between the smallest particles of matter. One can use transparencies to show force effects.

Material: New overhead transparencies, paper.

Procedure: Strongly rub a transparency with a piece of paper. Hang both transparency and paper close to each other. Rub two transparencies in a similar fashion on paper; bring the transparencies close to each other.

Observation: Initially paper and transparency are strongly attracted, the paper almost "sticks" to the transparency: objects attract if they are oppositely charged. In the second case, both transparencies repel each other: similarly charged objects repel each other.

E5.14 Ionic Bonding and Magnetic Model

Problem: Oppositely charged ions are attracted in rock salt crystal (Na^+ and Cl^-), similarly charged ions repel each other (Na^+ and Na^+, Cl^- and Cl^-). The attraction of all ions in salt crystals can be explained by the balance of attraction and repulsion forces. It is difficult to create such a model showing *electric* forces. It is, however, possible to implement a model for demonstration purposes with a balance of *magnetic* attraction and repulsion forces which thereby simulates a two-dimensional crystal.

Material: Round permanent magnets (10 mm), synthetic cover (5 mm for large discs and 2 mm for small discs), flat wooden box with see-through plexiglas.

Procedure: Place the large discs with their north pole facing upwards, they repel each other. Place the small discs with their north pole facing downwards, they also repel each other. Shake all discs in the closed box on the overhead projector, observe their arrangement.

Observation: Large and small discs attract each other. In the arrangement each large disc is surrounded by four small discs and each small disc surrounded by four large ones (see Fig. 5.19). This structural model can be compared to one layer of the sodium chloride structure. Similarly, salt structures gain stability from the balance of *electric* attraction and repulsion forces, in relation to the magnets there are magnetic forces.

Tip: If possible, separate the small and large magnets from each other before every new experiment: one kind of magnet on one side of the box and the other kind on the opposite side. After separating them, the experiment can be repeated with the same observation.

E5.15 Magnetic Force Fields

Problem: As we have already explained, it is not possible to see magnetic or electric forces. It is, however, possible to show these force effects: in the case of magnetic force, it can be shown using iron filings. In such a test model, it is possible to show non-directed forces which a magnet sends out in every direction. These non-directed force fields also surround every atom or ion – however, these are electric forces.

Material: Glass plate, magnet rod; iron filings (in a shaker).

Procedure: Cover the glass plate with a fine, see-through layer of iron filings. Gradually bring the rod magnet close under the glass plate (show it on the overhead projector). Cause the glass plate to vibrately using a metal object.

Observation: The filings arrange in a very specific manner (see Fig. 5.20): they show no specific direction, they are pointing in all directions around the magnet.

References

1. Roesky, H., Moeckel, K.: Chemische Kabinettstuecke. Weinheim 1996 (VCH)
2. Barke, H.-D.: Konzeption des strukturorientierten Chemieunterrichts. In: Barke, H.-D., Harsch, G.: Chemiedidaktik Heute. Heidelberg 2001 (Springer)
3. Wirbs, H.: Modellvorstellungen und Formelverstaendnis im Chemieunterricht. Dissertation. Muenster 2002 (Schueling)
4. Barke, H.-D., Sauermann, D.: Memorymetalle – sie besitzen ein Formgedaechtnis. Praxis Chemie 47 (1998), Heft 3, 7
5. Hilbing, C., Barke, H.-D.: Ionen und Ionenbindung: Fehlvorstellungen hausgemacht! Ergebnisse empirischer Erhebungen und unterrichtliche Konsequenzen. CHEMKON 11 (2004), 115

References

6. Barke, H.-D.: Chemiedidaktik zwischen Philosophie und Geschichte der Chemie. Frankfurt 1998 (Lang)
7. Barke, H.-D.: pH-neutral oder elektroneutral? Über Schuelervorstellungen zur Struktur von Salzen. MNU 43 (1990), 415
8. Barke, H.-D., Selenski, T., Sopandi, W.: Mineralwasser und Modellvorstellungen. PdN-ChiS 52 (2003), H.2, 15
9. Sauermann, D., Barke, H.-D.: Chemie für Quereinsteiger. Band 1: Strukturchemie und Teilchensystematik. Muenster 1997 (Schueling)
10. Christen, H.R., Baars, G.: Chemie. Frankfurt 1997 (Diesterweg)
11. Barke, H.-D.: Einfuehrung des Ionenbegriffs durch Experimente zur Gefrierpunktserniedigung. NiU-Chemie 23 (1992), 93
12. Strehle, N.: Der Ionenbegriff im Chemieunterricht. MNU 60 (2007), 45
13. Hilbing, C.: Alternative Schuelervorstellungen zum Aufbau der Salze als Ergebnis von Chemieunterricht. Dissertation. Muenster 2003 (Schueling)
14. Schwoeppe, C.: Fehlvorstellungen zum Ionenbegriff und Vorschlaege zu deren Korrektur. Staatsexamensarbeit. Muenster 2004 Taber, K.S.: Understanding Ionisation Energy: Physical, Chemical and Alternative Conceptions. CERP 4 (2003), 149
15. Taber, K.S.: Chemical Misconcetions – Prevention, Diagnosis and Cure. London 2002 (Royal Society of Chemistry)
16. Taber, K.S.: Building the structural concepts of chemistry: Some considerations from educational research. CERP 2 (2001), 123
17. Barker, V.: Beyond Appearances: Students' Misconceptions About Basic Chemical Ideas. London 2000 (Royal Society of Chemistry)
18. Sumfleth, E., Ploschke, B., Geisler, A.: Schuelervorstellungen und Unterrichtsgespraeche zum Thema Saeure-Base. Muenster 1999 (LIT)
19. Butts, B., Smith, R.: HSC chemistry students' understanding of the structure and properties of molecular and ionic compounds. Research in Science Education 17 (1987), 192
20. Boo, H.K.: Students' understanding of chemical bonds and the energetics of chemical reactions. Journal Research in Science Teaching 35 (1998), 569
21. Griffith, A.K., Preston, K.R.: Grade-12 students' misconceptions relating to fundamental characteristics of atoms and molecules. Journal Research in Science Teaching 29 (1992), 611
22. Harrison, A.G., Treagust, D.F.: Secondary students' mental models of atoms and molecules: Implications for teaching chemistry. Science Education 80 (1996), 509
23. Reiners, Ch.: Die chemische Bindung – Lernhindernisse und moegliche Lernhilfen. CHEMKON 10 (2003), 17
24. Taber, K.: The Truth About Ionic Bonding. Essex 1995
25. Huheey, J.E.: Anorganische Chemie – Prinzipien von Struktur und Reaktivität. Berlin 1988
26. Barke, H.-D., Hilbing, C.: Leserbrief zu Reiners, Ch.: Die chemische Bindung – Lernhindernisse und moegliche Lernhilfen. CHEMKON 10 (2003), 93
27. Reiners, Ch.: Leserbrief zu [27], CHEMKON 10 (2003), 152
28. Sauermann, D.: Leserbrief zu [28], CHEMKON 11 (2004), 75
29. Barke, H.-D.: Salze – Strukturvorstellungen und Symbole. NiU P/C 33 (1985), 169
30. Pfundt, H.: Urspruengliche Vorstellungen der Schueler für chemische Vorgaenge. NU 29 (1975) Mulford, D.R., Robinson, W.R.: An Inventory for Alternate Conceptions among First-Semester General Chemistry Students. Journal of Chemical Education 79 (2002), 739
31. Schmidt, H.J.: Stolpersteine im Chemieunterricht. Frankfurt 1990 (Diesterweg)
32. Taber, K.S.: Student conceptions of chemical bonding: using interviews to follow the development of A-level students' thinking. Conference Paper 1993
33. Strehle, N.: Die chemische Symbolsprache und deren Einfluss auf Einstellungen der Schueler und Schuelerinnen zum Chemieunterricht. Staatsexamensarbeit. Muenster 2002
36. Harsch, G.: Stereobilder zum Training des Raumvorstellungsvermoegens. In: Barke, H.-D., Harsch, G.: Chemiedidaktik Heute. Heidelberg 2001 (Springer)

37. Moeller, B., Barke, H.-D.: Interaktive Software zum Entdecken chemischer Strukturen. In: Zur Didaktik der Physik und Chemie. Probleme und Perspektiven. Berlin 2001 (Leuchtturm)
38. Moeller, B., Barke, H.-D.: Metalle und Salze. Computervisualisierungen von Strukturmodellen im Unterricht der Sekundarstufe I. Unterricht Chemie 16 (2005), Heft 89

Further Reading

Ben-Zvi, R., Eylon, B., Silberstein, J.: Students' visualisation of chemical reactions. Education in Chemistry 24 (1987), 117
Ben-Zvi, R., Eylon, B.S. Silberstein, J.: Is an atom of copper malleable? Journal of Chemical Education 63 (1986), 64
Gensler, W.: Physical versus chemical change. Journal of Chemical Education 47 (1970), 154
Kind, V.: Chemical concepts: chemical bonds. Education in Chemistry 40 (2003), 93
Peterson, R.F.: Tertiary students understanding of covalent bonding and structure concepts. Journal of Chemical Education 70 (1993), 11
Peterson, R.F., Treagust, D.F.: Grade-12 students' misconceptions of covalent bonding. Journal of Chemical Education 66 (1989), 459
Peterson, R.F., Treagust, D.F., Garnet, P.: Development and application of a diagnostic instrument to evaluate grade-11 and grade-12 students' understanding of covalent bonding and structure following a course of instruction. Journal Research in Science Teaching 26 (1989), 301
Strong, L. E.: Differentiating physical and chemical changes. Journal of Chemical Education 47 (1970), 689
Taber, K.S.: Misunderstanding the ionic bond. Education in chemistry 31 (1994), 100
Taber, K.S.: The secret life of the chemical bond: students' anthropomorphic and animistic references to bonding. International Journal of Science Education 18 (1996), 557
Tan, K.C., Treagust, D.F.: Evaluating students' understanding of chemical bonding. School Science Review 81 (1999), 294

Fig. 6.1 Concept cartoon concerning chemical equilibrium

Chapter 6
Chemical Equilibrium

In order to understand most of the basic concepts in chemistry, chemical equilibrium is enormously important. In this sense Berquist and Heikkinen [1] state: "Yet equilibrium is fundamental to student understanding of other chemical topics such as acid and base behavior, oxidation–reduction reactions, and solubility. Mastery of equilibrium facilitates the mastery of these other chemical concepts".

Unfortunately, it seems to be difficult to teach this topic. Finley, Stewart and Yarroch [2] studied the level of difficulty of various themes in chemistry and reported the results of 100 randomly chosen teachers of chemistry from Wisconsin who chose chemical equilibrium as being clearly the most difficult theme overall. Berquist and Heikkinen [1] noted in addition: "Equilibrium, considered one of the more difficult chemical concepts to teach, involves a high level of students' misunderstanding". One can therefore expect a large variety of misconceptions because of the difficulties in teaching this subject as well as for understanding it.

6.1 Overview of the Most Common Misconceptions

Tyson, Treagust and Bucat [3], Banerjee and Power [4], and Hackling and Garnett [5] studied students' comprehension of chemical equilibrium. The following misconceptions were discovered in these studies: "You cannot alter the amount of a solid in an equilibrium mixture; (...) the concentrations of all species in the reaction mixture are equal at equilibrium" [3]. "Large values of equilibrium constant imply a very fast reaction; (...) increasing the temperature of an exothermic reaction would decrease the rate of the forward reaction; (...) the Le Chatelier's principle could be used to predict the equilibrium constant" [4]. "The rate of the forward reaction increases with the time from the mixing of the reactants until equilibrium is established; (...) a simple arithmetic relationship exists between the concentrations of reactants and products at equilibrium (e.g. concentrations of reactants equals concentrations of products); (...) when a system is at equilibrium and a change is made in the conditions, the rate of the forward reaction increases but the rate of the other reaction decreases (...) the rate of the forward and reverse reactions could be affected differently by addition of a catalyst" [5].

Berquist and Heikkinen [1] have summarized students' misconceptions concerning chemical equilibrium in many areas as follows:

1. Students show confusion regarding amounts (mol) and concentration (mol/l) by attempting to compute concentrations when given molarity; expressing an uncertainty when to use volume; assuming stoichiometric mol ratios apply among product and reactant concentrations, assuming molar amounts are equal even when one is in excess.
2. Students show confusion over the appearance and disappearance of material by assuming concentrations fluctuate as equilibrium is established: a reaction is reversible yet goes to completion; that the forward reaction must be completed before the reverse one starts; and addition of more reactant changes only the product concentration.
3. Students show confusion over the meaning of K_c by describing it as varying in value while at constant temperature, assuming that the value changes with amounts of reactant products.
4. Students show confusion over the use of Le Chatelier's principle by attempting to adjust a system that is already at equilibrium; to change concentration of the added reactant only; to change concentration values of all species present except the added reactant; uncertainty of how a temperature, volume, or pressure change (including the addition of a non-reacting gas) will alter the equilibrium concentrations.

Kousatana and Tsaparlis [6] present another summary of misconceptions of various sub-themes regarding chemical equilibrium.

In 1992–1994, Kienast [7] carried out tests on chemical equilibrium with over 12,000 students in four test cycles. The following misconceptions, which were also described by the above-mentioned authors, were observed with particular regularity: "In equilibrium the sum of the amount of matter (concentration) of reactants is equal to the sum of the amount of matter (concentration) of the products; (...) in equilibrium the amounts (concentrations) of all substances which are involved in equilibrium are the same; (...) the sum of the amounts of matter (concentrations) remain the same during a reaction; (...) data which has been supplied on the amount of matter (concentration data) should be multiplied with stoichiometric coefficients from the reaction equation, in order to find the 'true' amount of matter concentration" [7].

6.2 Empirical Research

With the help of a questionnaire based on the above-mentioned inquiries, Tanja Osthues [8] shows typical misconceptions that occur with students in the advanced classes of some academic high schools around Münster in Germany.

6.2 Empirical Research

The results of the study are discussed based on the problems presented in the questionnaire.

The students were asked beforehand if they would be willing to fill out a questionnaire on a topic in chemistry without knowing what that exact theme would be. The teachers only saw the questionnaire on the day of the empirical study in order to avoid having the questions discussed in the lessons, only the school principal initially received a copy of the questionnaire.

At the beginning of the questionnaire there were some demographic questions. The age and gender of the student were first requested, then the type of school; the class and course type were to be filled in. In order to get an overview of the individual student's academic level in both Physics and Chemistry, they were asked for their previous grades in these subjects. It is of course not possible to get a perfect idea of the student's abilities merely by looking at their last course grade but it helps to clarify their general academic standing.

Because misconceptions about chemical equilibrium have been related in literature to student's lack of comprehension in basics of mathematics, they were also asked for their last grade in that subject. This way it was possible to gain an idea of whether there are correlations between students' misconceptions relating to chemical equilibrium and the quality of their grades in mathematics.

The questionnaire. It contains ten problems of which the first was chosen to include the specific distractors of the quoted test of Kienast [7]. The problems are individually introduced and when possible the percentage results of the choice of distractors are reflected using column diagrams. Some important requested explanations, which the students chose as their answers, are quoted word by word.

Problem 1 In a closed system, the following equilibrium can develop between the compounds ethane (C_2H_6), hydrogen (H_2) and ethene (C_2H_4):

$$C_2H_6(g) \rightleftharpoons C_2H_4(g) + H_2(g)$$

At the beginning of the reaction 8 mol C_2H_6 are present, at this time C_2H_4 and H_2 have not yet been formed. At equilibrium, 3 mol C_2H_4 are formed. How many mol of C_2H_6 and H_2 exist at equilibrium?

a) 2 mol C_2H_6 and 3 mol H_2
b) 3 mol C_2H_6 and 3 mol H_2
c) 4 mol C_2H_6 and 1 mol H_2
d) 5 mol C_2H_6 and 3 mol H_2
e) 6 mol C_2H_6 and 3 mol H_2

Please explain your answer in detail.

Answer a contains the false idea that "the sum of all amounts remains the same throughout a reaction" [7]. The students explain their answers as follows:

The total amount of material does not change; it should remain at 8 mol. Because 3 mol (of 8) are on the one side, there should be 5 on the other side.

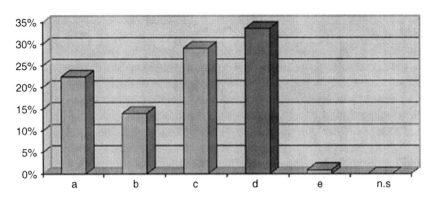

Problem 1 answer pattern (correct answer is d)

Answer a is correct, because 8 mol should be divided: 3 mol C_2H_4 exist after the reaction, C_2H_6 and H_2 should make up a total of 2 + 3 = 5.

When 3 mol C_2H_4 exist, 3 mol H_2 also exist. In order to attain the 8 mol C_2H_6, only 2 mol C_2H_6 can exist: 8 – 6 = 2.

In students' comments regarding **Answer b,** the expected incorrect answer pops up, i.e. that "in equilibrium, the amounts of matter are the same" [7]:

In each case, it has to be 3 mol.
We have an equilibrium when the same amount exists on either side.

Answer c is the most commonly chosen of the incorrect answers. The idea that "the total amount of matter remains the same during a reaction" and "at the same time, the total of the amounts of reactants is equal to the sum of the amounts of products" [7] can be proven in the student's comments:

4 mol exist on both sides of the reaction, therefore the reaction is in equilibrium, and in addition the total of the original matter is 8 mol.
On the side of the products and reactants there are 8 mol altogether. Because 3 mol C_2H_4 originate, there should be also 4 mol H_2 on the side of the products: 3 mol + 1 mol = 4 mol.
Equilibrium is reached when the same mass collects on both sides. From 8 mol we have on the right side 3 mol C_2H_4 and 1 mol H_2, accordingly on the reactant side 4 mol C_2H_6.
There must be 8 mol – in equilibrium there should be an equal number of mol on both sides.

The correct answer is **Answer d** – no incorrect explanation can be found.

Problem 2 In a closed system, one can create the following equilibrium between the substances carbon dioxide (CO_2), water (H_2O) and carbonic acid (H_2CO_3):

$$CO_2(aq) + H_2O(l) \rightleftharpoons H_2CO_3(aq)$$

At the beginning of the reaction 5 mol CO_2 and 5 mol H_2O are present. At this time, H_2CO_3 has not yet been formed. At equilibrium, 2 mol H_2CO_3 were formed. How many mol of CO_2 and H_2O exist at equilibrium?

6.2 Empirical Research

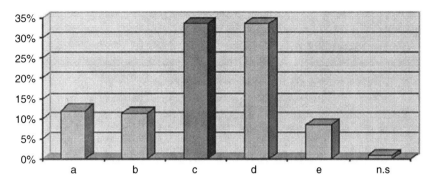

Problem 2 answer pattern (correct answer is c)

a) 1 mol CO_2 and 1 mol H_2O
b) 2 mol CO_2 and 2 mol H_2O
c) 3 mol CO_2 and 3 mol H_2O
d) 4 mol CO_2 and 4 mol H_2O
e) 3 mol CO_2 and 5 mol H_2O

Please explain your answer in detail.

In the student's commentaries for **Answer a**, the expected incorrect answer can be found, that "at equilibrium the sum of the concentrations of reactants is equal to those of products" [7]:

1 mol + 1 mol are 2 mol, on the other side we have 2 mol H_2CO_3 at equilibrium. The side of products should have the same amount as the side of reactants: 2 mol. We have an equilibrium when the same amount of matter exists on both sides.

Answer b is given because of the alternative concept, that "concentrations of all involved substances which are in equilibrium are equal" [7]. The students' explanations convey these misconceptions:

Because we have an equilibrium and 2 mol H_2CO_3 exist on the right side, the same 2 mol CO_2 should exist on the left side, consequently 2 mol H_2O exist.

Answer c is correct – this answer was chosen by approximately one third of the students, the explanations given in the students' commentaries are good.

Just as many students who chose the correct answer also chose **Answer d**. The misconception that "the sum of concentrations remains in a chemical reaction" is confirmed in the students' explanations:

I should have 10 mol. I assume that from the 5 mol H_2O and 5 mol CO_2 an equal amount of each transfer over to 2 mol H_2CO_3. This would mean that each one remains the same: 4 mol H_2O and 4 mol CO_2.

At the beginning, 10 mol and 2 mol are formed, 8 mol are missing, therefore Answer d. If 5 mol H_2O and 5 mol CO_2 react and 2 mol H_2CO_3 are formed, then 4 mol H_2O and 4 mol CO_2 have to react in order to attain a total of 10 mol.

In order to re-establish the equilibrium, 8 mol are missing. Because an equal amount of H_2O and CO_2 existed at the beginning, the same amount should exist at the end: 4 mol and 4 mol.

Again, here, one should come up with 1 mol. If 2 mol H_2CO_3 are existing, then 4 mol H_2O and 4 mol CO_2 have yet to be formed.

Behind **Answer e** the misconceptions lies: "the sum of moles remains the same throughout a chemical reaction" [7]. This misconception however, cannot be verified in the students' commentaries.

Problem 3 The following equilibrium can be found between the compounds nitrogen dioxide (NO_2) and dinitrogen tetraoxide (N_2O_4):

$$2NO_2(g) \leftrightarrows N_2O_4(g)$$

At the beginning of the reaction 7 mol NO_2 were present. At this time N_2O_4 had not yet formed. At equilibrium 2 mol N_2O_4 are formed. How many mol NO_2 exist at equilibrium?

a) 1 mol NO_2
b) 2 mol NO_2
c) 3 mol NO_2
d) 5 mol NO_2
e) 6 mol NO_2

Please explain your answers in detail.

Answer a leads to the conception, "that at equilibrium the sum of the concentrations of reactants is equal to the sum of the concentrations of products, and at the same time the given concentrations have to be multiplied with the stoichiometric coefficients from the reaction, in order to get a true concentration" [7]. The student's explanations substantiate this idea:

Because NO_2 is half of N_2O_4, the molar mass should be half: 2 mol/2 = 1 mol. Adjustment $2 \cdot 1$ mol $\leftrightarrow$ 2 mol.

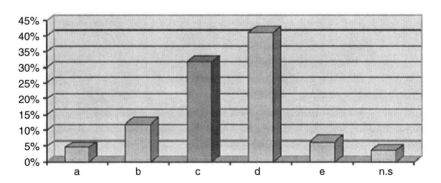

Problem 3 answer pattern (correct answer is c)

6.2 Empirical Research

At equilibrium, we have 1 mol NO_2, because there are 2 mol NO_2 molecules therefore 2 mol exist on either side.

Many students check **Answer b** as being correct and, again, exhibit the expected misconception in their explanation, i.e. "at equilibrium, the concentrations of all involved substances are equal" [7]:

In order to maintain the equilibrium, it must also be correct within the equation. If 2 mol N_2O_4 exist, one also needs 4 N molecules for the reactant; this can be attained if one has 2 mol NO_2 for the reactant.

Because there are 2 mol N_2O_4 i.e. 4 N atoms are formed on the right side, therefore 4 N atoms, i.e. 2 mol NO_2 should also be on the left side.

The correct solution lies in **Answer c,** in each case it is explained correctly.

"The idea, that the sum of concentrations remains constant in a chemical reaction" [7], leads to the choice of **Answer d** which was the most common answer amongst the students.

Problem 4 In an enclosed system, the following equilibrium can exist between the compounds hydrogen (H_2), iodine (I_2) and hydrogen iodide (HI):

$$2HI(g) \leftrightarrows H_2(g) + I_2(g)$$

At the beginning of the reaction 6 mol HI are introduced. At this time, H_2 and I_2 have not yet formed. At equilibrium, 1 mol H_2 exist. How much HI and I_2 exist at equilibrium?

a) 1 mol HI and 1 mol I_2
b) 2 mol HI and 1 mol I_2
c) 3 mol HI and 2 mol I_2
d) 4 mol HI and 1 mol I_2

Please explain your answers in detail.

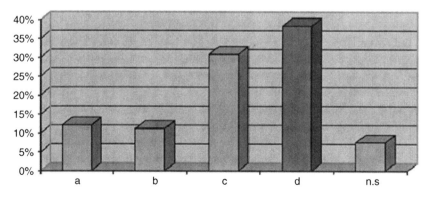

Problem 4 answer pattern (correct answer is d)

The answers and explanations contain misconceptions as already discussed. The following problems convey a connection to chemical equilibrium and – without giving any choice of answers – students were requested to supply answers, if possible detailed explanations.

Problem 5 In the Haber–Bosch process for the preparation of ammonia, the following reactions occur in a closed system:

$$N_2(g) + 3H_2(g) \leftrightarrows 2NH_3(g)$$

After equilibrium has been achieved, the pressure is increased at constant temperature. What happens to the system? Please explain your answer in detail.

The correct answer is that equilibrium moves to the product side because the amount of molecules are halved which leads to an "evasion of the pressure". The majority of the students recognized this correctly and gave a correct explanation. Some of the students did not think that the pressure increase causes "any change." They gave the following explanations:

There would be no changes in equilibrium because it is in balance; it does not change anymore. Nevertheless, an exchange occurs between both sides.
Because it is a closed system, equilibrium is not altered.

The emphasis on a closed system seems to cause the idea, that no change in equilibrium whatsoever is possible, not even through a change in pressure. The next problem does not deal with closed systems, but with attractive color changes by special reactions – depending on the change of the equilibrium.

Problem 6 Observe the following reversible reaction, which appears blue in equilibrium:

$$[Co(H_2O)_6]^{2+}(aq, \text{ pink}) + 4\ Cl^-(aq) \leftrightarrows [CoCl_4]^{2-}(aq, \text{ blue}) + 6H_2O(l)$$

What happens when water is added to this system?
Please explain your answer in detail.

As expected, a majority of the students answered this question correctly and described the color change to pink in favor of the reactants. However, many of the explanations of the correctly answered questions are contestable because a spatial separation of the substances in equilibrium exists in the imagination of students (bold markings by the authors):

*When water is added to this system, an excess exists on the **right side**. Because this*
    *is balanced, more substances are formed on the **left side**.*
*The equilibrium shifts **towards the left**. The additional water causes a rearrangement*
    of equilibrium. In order to regain the same molar amounts on either side, the
    equilibrium must be shifted more towards the left.
According to Le Chateliers principle, the solution turns pink because the added
    *substance is used up **on the right side**.*

6.2 Empirical Research

Equilibrium moves more towards the **left side** *because of the increase of water concentration on* **the right side, where the blue solution is**.

Several of the students are of the opinion that through the addition of water, a dilution of the blue color takes place. They give the following reasons:

The color is diluted, every color is diluted though the addition of water.
The colors become brighter because the substances are diluted, however, their effectiveness remains the same.
Because the solution is blue in equilibrium it can, at the most, through the addition of water, change in such a way that the blue color becomes weaker because all is diluted. It does not lose its equilibrium and therefore the pink color cannot predominate.

There are other wrong answers and related misconceptions, i.e. that the addition of water has no influence on the equilibrium reaction:

Nothing happens because water does not change the reaction.
Water does not affect the reaction in any way.

It appears to be difficult for students to differentiate between the solvent water and an additional partner in the reaction. These differences should be clearly pointed out to students who have to appropriately judge a complex reaction according to the pattern in Problem 6. The following reaction is about water as a reaction partner and as a solvent.

Problem 7 In a freshly produced 0.5 M solution of sodium dichromate ($Na_2Cr_2O_7$), the following equilibrium sets in after a certain time:

$$2CrO_4^{2-} \text{ (aq, yellow)} + 2H^+(aq) \leftrightarrows Cr_2O_7^{2-}(aq, \text{ red}) + H_2O(l)$$

What happens if one adds 10 ml of a 0.5 M solution of sodium dichromate to 10 ml of the solution mentioned above?
    Please explain your answers in detail.

A mere tenth of the questioned students correctly recognized that nothing changes in the equilibrium, because the same 0.5 molar solutions are added, i.e. the same concentration of particles are available in both solutions and no changes in concentration can be found.

Other students came to false conclusions: the equilibrium would move to the reactant side. Several explanations point to that:

If one adds sodium dichromate, there is an excess on the right side. Therefore, more substances are formed on the left side whereas the concentrations decrease on the right side.
Equilibrium moves over to the left because with more $Na_2Cr_2O_7$, more $Cr_2O_7^{2-}$ is added and can react with water to more CrO_4^{2-} and H^+.
If one adds sodium dichromate, the concentration of $Cr_2O_7^{2-}$ increases. According to Le Chatelier's principle, the equilibrium moves to the left: the solution turns yellow.

It appears that they utilize their knowledge regarding Le Chatelier's principle – but without regarding the mentioned conditions. It is not clear if a misconception lies behind these answers – even reading over the questions too superficially could lead to the incorrect answers.

The next problem will show the same difficulty with the unreflected usage of Le Chatelier's principle: students should state whether the addition of a solid substance to a mixture of matter, which is in equilibrium, leads to the change of concentrations or not.

Problem 8 Solid calcium carbonate ($CaCO_3$) exposed to intense heat forms calcium oxide (CaO) and carbon dioxide (CO_2), until this equilibrium is reached:

$$CaCO_3(s) \leftrightarrows CaO(s) + CO_2(g)$$

What happens to the concentration of CO_2 if solid calcium oxide is added to the equilibrium?
Please explain your answers in detail.

The correct answer that the equilibrium does not change through the addition of a solid substance was only recognized by one tenth of the students. The given commentaries exclusively supplied correct explanations.

However, many students used Le Chatelier's principle without exactly contemplating, in this particular case, that "no stress" is being put on this system in equilibrium. Most of the students guessed that there is a decrease in CO_2 and therefore a shifting of the equilibrium on the side of the reactants. They explained their assumption as follows:

The concentration of CO_2 decreases, because this reacts with the added CaO to produce $CaCO_3$: equilibrium therefore shifts to the left.
Equilibrium is disturbed because more calcium oxide would be present: the concentration of CO_2 decreases.
The concentration lessens, because CO_2 reacts with CaO – the equilibrium shifts to the left.
The concentration of CO_2 regresses, because CaO is in excess and therefore it reacts with CO_2 to $CaCO_3$. CO_2 is therefore used which leads to a decrease in concentration.
The concentration of CO_2 would be lowered, because it is used together with calcium oxide.
If the CaO amount is increased, then the amount of CO_2 has to be smaller in order to retain equilibrium.
The concentration of CO_2 sinks because an excess of CaO benefits the reaction to the left.

There are students who remember that the addition of a solid substance does not disturb a reaction in equilibrium. They think, however, that calcium oxide dissolves after some time and has an influence on the equilibrium reaction:

Solid calcium oxide has to dissolve first in order to be able to react. Then, the equilibrium is shifted on the side on which the added substance is used.
More $CaCO_3$ is formed, because calcium oxide can be dissolved and can influence the CO_2.

6.2 Empirical Research

In Problem 3, a survey was carried out on the equilibrium "2 $NO_2(g) \leftrightarrows N_2O_4(g)$". In order to find out and study concepts regarding the influence of concentrations. In Problem 9, the same equilibrium is used in order to discuss the influence of temperature.

Problem 9 The reaction of nitrogen dioxide (NO_2) to dinitrogen tetraoxide (N_2O_4) is exothermic:

$$2NO_2(g, \text{ brown}) \leftrightarrows N_2O_4 \text{ (g, colorless); forward reaction exothermic}$$

After equilibrium has been established, the temperature is raised at constant pressure. What will happen to the system? Please explain your answers in detail.

Most students, as expected, answered the problem correctly: the equilibrium avoids the high temperature by realizing the endothermic reverse reaction; a new equilibrium position in favor of the reactants is formed.

Some students stated that the equilibrium would shift to the side of the products. They gave following explanations:

The equilibrium shifts towards the right so that more heat can be released.
The equilibrium shifts towards the right, due to the exothermic reaction and additional increase in temperature.

The last Problem concerns the influence of an increase in temperature to an endothermic reaction already at equilibrium.

Problem 10 The reaction of carbon (C) with carbon dioxide (CO_2) to form carbon monoxide (CO) is endothermic:

$$C(s) + CO_2(g) \leftrightarrows 2CO(g); \text{ forward reaction endothermic}$$

After equilibrium has set in, one increases the temperature at constant pressure. What will happen to the system? Please explain your answer in detail.

Similar to Answer 9, there are many correct answers and explanations which suggest the formation of additional carbon monoxide. However, the argument of a shifting of equilibrium to the reactant side is also used and explained as follows:

The exothermic reaction (formation of C + CO_2) is boosted here. Therefore more of these substances are formed. Equilibrium would be disturbed.
One must supply additional energy so that carbon monoxide can be formed. If the temperature is increased (more energy is supplied), then the carbon monoxide would have a lot more energy. I could imagine that CO reacts into C and CO_2 because the reaction is reversible.

It is apparent in the last answer that the terms "temperature" and "energy" are not being appropriately differentiated which results in misconceptions. In Chap. 10, these questions will be evaluated and suggestions will be made towards improving lessons.

6.3 Teaching and Learning Suggestions

It is not simply and convincingly possible to counter the cited common misconceptions, i.e. that chemical equilibrium show "equal concentrations of reactants and products". As an introduction, it might be helpful to show melting equilibria with different ice–water mixtures, or solubility equilibria with different amounts of the solid.

Melting equilibrium. Different ice–water mixtures are produced, stirred for some time, and measured with the thermometer (see E6.1): the temperatures are always 0°C. The following melting equilibrium exists:

$$\text{ice (s, 0°C)} \leftrightarrows \text{water (l, 0°C)}$$

The melting equilibrium does not depend on the amount of ice or water, but it exists in each ice–water mixture. If one heats the mixture then a part of the ice melts. When the mixture is still stirred, the temperature of 0°C stays constant (see E6.1): the energy is used to separate a special amount of water molecules from ice crystals to form water. Similarly, the evaporation equilibrium in a closed system under vacuum could be discussed (see Fig. 6.2): water molecules leave the liquid phase and enter into the gas phase (steam); in the same manner, other molecules move from the steam into the liquid phase. Even in this case, the amount of liquid does not make any difference.

Solubility Equilibrium. The misconceptions regarding the amount of solid materials in equilibrium and the dynamic aspect are equally important in the discussion. If one observes a saturated sodium chloride solution together with solid sodium chloride, and adds an additional portion of solid sodium chloride to it, this portion sinks down without dissolving (see E6.2). If one measures the density of the saturated solution before and after the addition of salt portions, one gets the same measurements (see E6.2). The concentration of the saturated solution does not depend on how much solid residue is present; equilibrium sets in between the saturated solution and arbitrary amounts of solid residue (see Fig. 6.2):

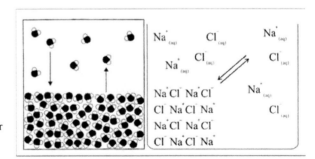

Fig. 6.2 Mental models for evaporation and solubility equilibrium [9]

6.3 Teaching and Learning Suggestions

$$Na^+Cl^-(s, \text{ white}) \leftrightarrows Na^+(aq) + Cl^-(aq)$$

There exists one kind of sodium iodide, which emits radiation: it contains radioactive isotopes of iodide ions. If one adds a small sample of that radioactive salt to a saturated solution of sodium iodide, then it sinks and a certain level of radioactivity can be determined at the base of the beaker using a Geiger counter. If one carries out the same measurement a few days later, then the radiation can also be determined in the solution, although the same amount of the solid is observed: sodium ions and iodide ions have transferred over from the solid to the saturated solution and other ions have transferred over from the solution to the solid salt. Therefore, chemical equilibria are not static but rather dynamic: back and forth reactions are constantly happening at an equal rate.

One can also observe the surface of salt crystals on the bottom of the saturated solution using a magnifying glass or by taking photographs over a long period: several crystals constantly increase in size, whereas others get smaller. A dynamic exchange of matter goes on between the solid residues and the saturated solution, forward and reverse reactions are constantly occurring: a dynamic equilibrium exists.

One cannot see constant reactions from saturated salt solution to solid salt and back. In order to have a better idea of dynamic equilibrium, it is of course possible to revert to a model experiment (see E6.3). Two similar measuring cylinders are prepared, 50 ml of water are placed in one of the cylinders, and the other one remains empty (see Fig. 6.3). Using two glass tubes of equal diameter to transport water back and forth, water is continuously transported between the two cylinders: after several such exchanges, 25 ml of water remains in each of the cylinders, the water level does not change despite carrying constant volumes of water back and forth. If a similar experiment is done using two glass tubes with different diameters, then one cylinder would perhaps have the volume of 10 ml and the other would have 40 ml "in equilibrium": the water level does not change although the same amount of water is continuously carried back and forth in the two different glass tubes (see Fig. 6.3).

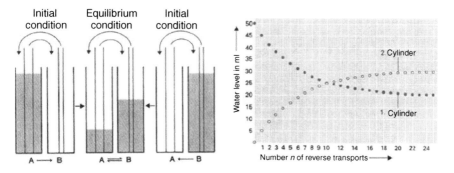

Fig. 6.3 Model experiment for the dynamic aspect of the equilibrium [10]

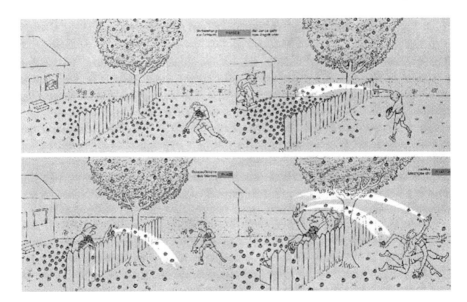

Fig. 6.4 An "apple fight" as a mental model for dynamic equilibrium [11]

An "apple fight" between an old man and the boy next-door [11] carries out a different model experiment on dynamic equilibrium. The boy is supposed to discard bad apples and just throws them into his neighbor's garden. The neighbor reacts and throws them back – his own garden is already full of apples ("large concentration"). Whereas the boy has to really hurry in order to collect his bad apples ("small concentration"), the old man effortlessly collects the same amount of apples, which he throws back. Finally, in balance, six apples are thrown in and six are returned (see Fig. 6.4) – despite different concentrations of apples on both sides.

If concentrated hydrochloric acid is placed in a clear saturated sodium chloride solution, then white sodium chloride precipitates as fine crystals (see E6.4). They "trickle" into the glass from above to below and fall to the bottom. The drastic increase in concentration of Cl^-(aq) ions causes a disturbance in equilibrium, and so much sodium chloride precipitates until a new equilibrium with higher concentrations of Cl^-(aq) ions than of Na^+ ions sets in (see Fig. 6.5). The new saturated

Fig. 6.5 Mental model on reaction of saturated salt solution with hydrochloric acid

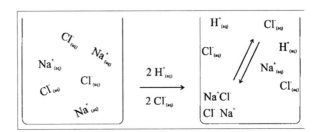

6.3 Teaching and Learning Suggestions

solution contains, apart from the salt ions of course, H^+(aq) ions of hydrochloric acid. This principle of "getting rid of stress" is also known as Le Chatelier's principle: the dynamic chemical equilibrium evades the stress of greater concentrations of Cl^-(aq) ions by forming solid sodium chloride (see Fig. 6.5).

Solubility Product. If calcium sulfate powder (gypsum) is mixed well with water and the suspension is left to stand, a white solid sinks down to the bottom (see E6.5). The question arising from the amount of solid substance is whether a part of the calcium sulfate dissolves or if the substance is insoluble. Testing the electrical conductivity (see E6.5), however, shows a much higher value than with distilled water: calcium sulfate dissolves in very minute amounts; a dynamic equilibrium is formed between the solid residue and the saturated solution:

$$Ca^{2+}SO_4^{2-}(s, \text{ white}) \leftrightarrows Ca^{2+}(aq) + SO_4^{2-}(aq)$$

Magnesium sulfate and calcium sulfate solutions of equal concentrations show approximately the same electrical conductivity [9]. If one compares electrical conductivity of the saturated calcium sulfate solution with the conductivity of various standard solutions of magnesium sulfate (see E6.5), one finds consistency in the concentration of the saturated solution:

$$c(\text{calcium sulfate}) = 10^{-2} \text{ mol/l}$$

Accordingly, for the pure saturated calcium sulfate solution we know (see Fig. 6.6):

$$c(CaSO_4) = 10^{-2} \text{ mol/l}, \text{i.e. } c(Ca^{2+}) = 10^{-2} \text{ mol/l} \quad c(SO_4^{2-}) = 10^{-2} \text{ mol/l}$$

The solubility product expression is defined as:

$$K_{sp}(CaSO_4) = c(Ca^{2+}) \times c(SO_4^{2-}) = 10^{-4}$$

If one is dealing with a diluted calcium sulfate solution, saturation can be attained in three different ways (see Point A in Fig. 6.6): One continues to

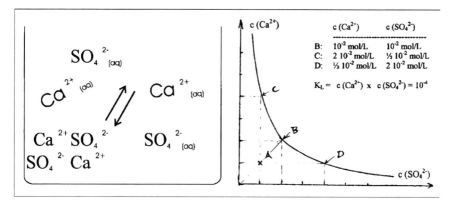

Fig. 6.6 Mental model on the solubility equilibrium of calcium sulfate

add solid calcium sulfate and reaches saturation (Point B). It is however also possible to add dropwise concentrated calcium chloride solution, thereby increasing the concentration of Ca^{2+}(aq) ions until the first calcium sulfate crystals precipitate (Point C). It is also possible to add some sodium sulfate solution, thereby increasing the concentration of SO_4^{2-}(aq) ions, until the first solid calcium sulfate precipitates (Point D). In each case, we have a pair of values for the saturation equilibrium on the hyperbolic curve (see Table in Fig. 6.6), these pairs follow the solubility product.

If one varies concentrations of ions involved in equilibrium by adding same kind of ions, then it is obvious that the product of ion concentrations is always constant, that this product has always, at constant temperature, the value $K_{sp} = 10^{-4}$. Tables and hyperbolic figures may demonstrate the concentration dependence of the related ions (see Fig. 6.6). It is, of course, common in many tables and textbooks to show this value with the units of mol^2/l^2; the specific scientific derivation is a dimensionless absolute.

The solubility equilibrium of calcium sulfate can also be demonstrated by supplementing portions of sodium sulfate and calcium chloride solutions (see E6.5): using highly concentrated solutions solid calcium sulfate precipitates. In addition to Ca^{2+}(aq) ions and SO_4^{2-}(aq) ions, the solution also contains Na^+(aq) ions and Cl^-(aq) ions:

$$Ca^{2+}(aq) + SO_4^{2-}(aq) \leftrightarrows Ca^{2+}SO_4^{2-}(s, \text{ white})$$

Solubility equilibrium sets in using solid matter as well as solutions, it can be approached from both sides: model experiments should be used to demonstrate this (see Figs. 6.2 and 6.3).

Boudouard Equilibrium. Equilibria can be well demonstrated when gases are involved. Carbon is used to reduce iron oxide minerals for the production of iron metal in a blast furnace. However, carbon itself barely functions as a means of reduction but rather the carbon monoxide, CO. It results in a reaction of gaseous carbon dioxide with glowing carbon by a temperature of approximately 1000°C:

$$C(s, \text{ glowing}) + CO_2(g) \leftrightarrows 2CO(g)$$

In the experiment, one takes 50 ml of carbon dioxide in one syringe, granular carbon in a combustion tube, and on the other side an empty syringe (see E6.6). Using the hot flame of the burner, the carbon is heated until glowing; the carbon dioxide is pressed slowly over the glowing carbon. The gas volume increases when the carbon dioxide gas reacts with the glowing pieces of carbon. When the apparatus has cooled down, one measures 60 ml of a colorless gas after the reaction. Using a gas tracing instrument, i.e. Draeger tubes for CO, the presence of carbon monoxide can be easily detected; the presence of carbon dioxide is observed using limewater.

In order to clearly demonstrate the existence of equilibrium, one can first compare the volumes of the gas before and after the complete reaction and then, at equilibrium:

6.3 Teaching and Learning Suggestions

$$C(s, \text{ glowing}) + CO_2(g) \leftrightarrows 2\ CO(g)$$

before the reaction:	50 ml	0 ml
after complete reaction:	0 ml	100 ml
at equilibrium:	(50 – x) ml	2x ml
	40 ml	20 ml

Because 50 ml CO_2 theoretically delivers 100 ml pure CO in the complete reaction, both gases must be mixed in order to get the volume of 60 ml, namely (50 – x) ml CO_2 and (2x) ml CO. If one solves the equation "(50 – x) + 2x = 60 ml", one gets x = 10 ml: in equilibrium, we have 40 ml CO_2 and 20 ml CO i.e. 66% by volume CO_2 and 33% by volume CO.

If one refers to the temperature–volume diagram of the Boudouard equilibrium (see Fig. 6.7), one recognizes that the attained equilibrium exists at about 650°C, a temperature that can easily be produced with a good burner. One also recognizes that, at higher temperatures, the amount of carbon monoxide at equilibrium increases a lot and that the position of the equilibrium varies with the temperature.

Equilibrium Constant. The position of the CO_2/CO equilibrium can be explained with the help of the equilibrium constant, using either the partial pressures or concentrations of both gases. In none of the cases the solid carbon is considered in the Boudouard equilibrium: it neither adds to the complete pressure, nor does it allow a concentration to be specified.

For kinetic reasons, it follows that for equilibrium constants, the CO partial pressure or CO concentration should be used with the exponent of 2, partial pressure or concentration of CO_2 with the exponent of 1:

$$C(s) + CO_2(g) \leftrightarrows 2CO(g);\quad p_{CO}^2/p_{CO2} = K_P;\quad c_{CO}^2/c_{CO2} = K_c$$

For p_{total} = 1 bar (about 1 atm) at 650°C, we arrive at:

p_{CO} – 0.33 bar and p_{CO2} = 0.66 bar, it follows: $K_P = p_{CO}^2/p_{CO2} - (0.33)^2/0.66 - 0.17$

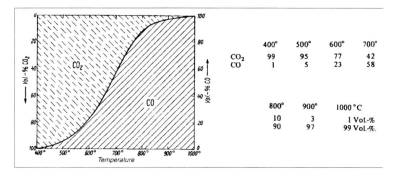

Fig. 6.7 Temperature–volume diagram of the Boudouard equilibrium [12]

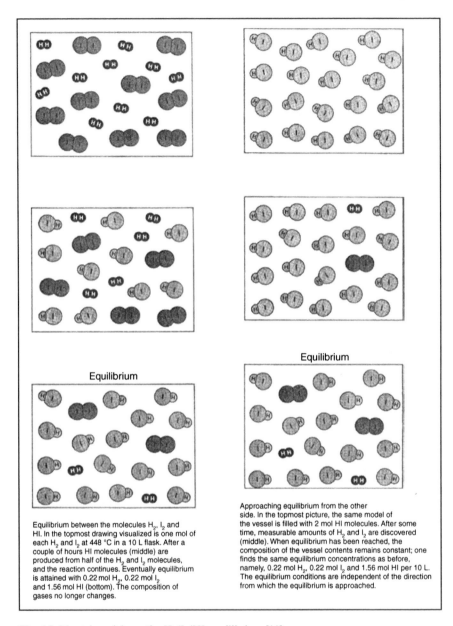

Fig. 6.8 Mental models on the $H_2/I_2/HI$ equilibrium [11]

Iodine–Hydrogen Equilibrium. This example should elucidate, using model drawings, that equilibrium sets in from both directions (see Fig. 6.8). It does not matter if one assumes an equimolar mix of iodine vapor and hydrogen gas, or one of pure hydrogen iodide gas – in both cases, one reaches an identical

6.3 Teaching and Learning Suggestions

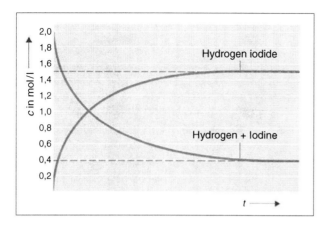

Fig. 6.9 Diagram for setting the $H_2/I_2/HI$ equilibrium from both sides [9]

equilibrium mixture. If one fills a 10 l flask with 1 mol H_2 molecules and 1 mol I_2 molecules and heats it to a temperature of 448°C, then one gets a mixture of 1.56 mol HI molecules, 0.22 mol H_2 molecules and 0.22 mol I_2 molecules (see Fig. 6.8). If one fills the same flask, at the same temperature, with 2 mol HI molecules, one achieves identical equilibrium concentrations. The experimental curves (see Fig. 6.9) give the same result and can be graphically shown with model experiments (see Fig. 6.3):

$$H_2(g) + I_2(g) \leftrightharpoons 2HI(g) \quad H_2(g) + I_2(g) \leftrightharpoons 2\ HI(g)$$

before the reaction:	1 mol	1 mol	0 mol	0 mol	0 mol	2 mol
at equilibrium:	0.22	0.22	1.56	0.22	0.22	1.56

If one uses the same amounts per 10 l, the concentrations in mol per liter are:

$$c(H_2) = 0.022 \text{ mol/l}, \ c(I_2) = 0.022 \text{ mol/l}, \ c(HI) = 0.156 \text{ mol/l}$$

The equilibrium constant follows:

$$K_C = c(HI)^2 / c(H_2) \times c(I_2) = 50.3$$

Ammonia Equilibrium. In school experiments, it is possible to quantitatively decompose ammonia gas with the help of a nickel catalyst into its components. From 50 ml ammonia, we can get 100 ml gas, namely 25 ml nitrogen and 75 ml hydrogen (see E6.7). However, it is not possible to reverse the procedure under normal pressure, i.e. to achieve the ammonia synthesis from the elements. The volume–temperature diagram (see Fig. 6.10) confirms this: one does not get any noticeable traces of ammonia at a pressure of 1 bar. In contrast, ammonia can almost be completely produced from the deployed gases at 200°C and a pressure of 1000 bar:

$N_2(g) + 3H_2(g) \leftrightarrows 2NH_3(g)$; forward reaction is exothermic

Now it is possible to discuss the influence of temperature and pressure on the position of equilibrium (see Fig. 6.10). The exothermic reaction of the elements requires a relatively low temperature. Since we get in the forward reaction from 4 mol molecules only 2 mol molecules, the formation of NH_3 molecules is preferred at high pressure.

This connection can be clearly visualized by a special model experiment (see Fig. 6.10): 12 molecules can be counted before applying pressure, only 11 molecules afterwards [13]. For economic reasons, industry uses medium temperatures of 450°C and pressures of around 300 bar: yields of approximately 40% volume of ammonia are large enough to continuously remove the ammonia from equilibrium thereby optimizing the production process.

Summary. The discussed equilibria show, in the cases in which the concentrations of the reaction partners are measured, that these concentrations are not inevitably the same as students assume (see Sect. 6.1). For the Boudouard equilibrium, completely different pairs of values are possible (see Fig. 6.7). The same applies for the hydrogen–iodine example (see Figs. 6.8 and 6.9) and for the ammonia example (see Fig. 6.10).

The Boudouard equilibrium also shows that the excess of solid matter does not play any role: how much carbon is used or which part of the carbon pieces are heated is irrelevant. Even in the furnace process, neither the amount of coal mixed into the iron ore nor how full the furnace is affects the position of the equilibrium. In order to have an optimal reaction, though, it is only necessary to have an excess of coal.

The same experiment also indicates, that reaction chambers "left and right" of equilibrium do not actually exist. Both of the involved gases and the carbon pieces are mixed all the time, neither "C and CO_2 are on the left side" nor is the "CO on the right side of equilibrium".

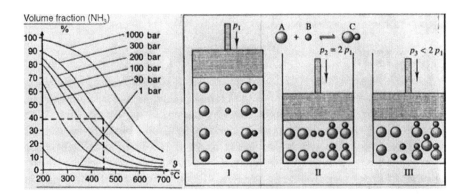

Fig. 6.10 Mental models of the ammonia equilibrium [13]

The students could, to some degree, see the solubility equilibrium with solid residue in equilibrium with separate substance phases: the solid residue is inevitably always at the bottom, the saturated solution on top. Perhaps it would be an advantage to not only present these equilibria with a solid *residue*, but also show that the solid matter, using a magnetic stirrer, *spreads* throughout the beaker.

Finally, influences of these parameters on the position of equilibrium could be discussed and looked at with temperature–volume diagrams (see Fig. 6.7) and with temperature–pressure diagrams (see Fig. 6.10). The model drawings on dynamic chemical equilibria and on particle processes (see Figs. 6.2, 6.5, 6.6, 6.8 and 6.10) appear to be extremely important for students' understanding. In this way, they do not have to learn hard-to-understand formulas and equations by heart, but with such drawings they would take note of the existence and number of affected atoms, ions and molecules. If they complete these model drawings themselves as homework or by copying from the blackboard, the learning success would be optimal.

We shall study the donor–acceptor reactions concerning chemical equilibrium in the next chapters: correlating drawings and mental models will be provided for better comprehension. This way, comprehension of chemical equilibrium can be intensified and completed.

6.4 Experiments on Chemical Equilibrium

E6.1 Melting Equilibrium

Problem: A widely accepted misconception on chemical equilibrium assumes that the concentration of reactants and products are equal. Attempts are made to demonstrate with different mixtures of varying ice and water portions, that at melting equilibrium, arbitrary amounts of water and ice can exist, that melting temperature of 0°C is always measured at normal conditions. If ice–water mixtures are heated, the energy is taken to separate water molecules from the ice lattice, not to increase temperatures: 0°C stays constant.

Material: Two beakers, thermometer, tripod and wire gauze; ice cubes.

Procedure: Fill a beaker half with water and add some ice cubes; fill another beaker half with water and as many ice cubes as possible. Mix the contents of the beakers stirring with thermometer until constant temperature is reached; and record the temperature. Heat the second beaker for some time with the help of a burner while stirring. Measure the temperature again.

Observation: Both the mixture with little ice and the mixture with a lot of ice show a constant temperature of 0°C. Even after heating the ice–water mixture, the temperature of 0°C exists regardless of the amounts of ice and water present.

E6.2 Solubility Equilibrium of Salt

Problem: Students often think that solubility equilibria depend on the amount of solid residue, or they succumb to the fallacy, that in the law of mass action must include the concentration of solid material. It should be shown, that it makes no difference how much solid residue is at equilibrium in the saturated salt solution.

Material: Large test tubes, hydrometer; sodium chloride, saturated sodium chloride solution.

Procedure: Fill a test tube two-thirds with saturated sodium chloride and measure the density with hydrometer. Add a spoon of sodium chloride, stir and measure the density again. Add several spoons of the salt to the solution, stir the solution and repeat the density measurement.

Observation: The density at room temperature is approximately 1.2 g/ml. It stays constant, even when small or large portions of salt are added. Although the salt portions swirl around during the stirring swirl process, they do not dissolve, but rather form a deposit.

Tip: The solid residue deposit might give pupils the false impression that substances in equilibrium "left" or "right" exist separately from each other as depicted in the reaction equations. The stirring or swirling of the solid residue with the saturated solution could be used to point out that "reactants and products" are not separated from each other but exist in a mixed state.

E6.3 Model Experiment on Dynamic Equilibrium

Problem: The dynamics in melting equilibria or solubility equilibria cannot be seen. It is possible, however, to carry out model experiments that demonstrate dynamic equilibrium. One takes two graduated cylinders, water as a model for involved matter, and transfers portions of water by glass tubes in both directions – as a model for the forward and reverse reaction (see also Fig. 6.3). When the volumes of water no longer change, then equilibrium has been reached – although the transportation of same amounts of water goes on to show a model of the dynamic equilibrium. One could also use the "apple fight" as a model, in order to demonstrate the dynamic equilibrium (see Fig. 6.4).

Material: Two identical 50 ml graduated cylinders, four 30 cm long glass tubes, two of 8 mm and two of 6 mm diameter.

Procedure: Fill one graduated cylinder with 50 ml of water (it models the amount of substance before equilibrium sets in), the other cylinder remains empty (no product initially). Hold two identical glass tubes (8 mm) in both hands, dip into both cylinders, close with index finger, lift out and transport the portion of water into the other cylinder. Repeat this procedure as many times as necessary until the water volumes in both cylinders no longer change. Repeat this experiment with glass tubes of different diameters.

Observation: At the end of the first test series, 25 ml of water is noted in each cylinder, the glass tubes can continue to transport water: the 25 ml level remains the same. At the end of the second test series, 30 ml water is placed in one cylinder, in the other one 20 ml are measured, several repetitions of this transportation of water do not change the volumes of water.

Tip: The experiment shows not only a model of the dynamic forward and reverse reactions of a system in equilibrium but also the fact that, at equilibrium, one can have different amounts of matter, not always 50:50. Depending on conditions (for the model the diameters of both glass tubes), all other values should also be taken into account: 40:60, 20:80, 10:90, etc.

E6.4 Solubility Equilibrium of Sodium Chloride

Problem: Solubility equilibrium of a salt is not limited to the concentrations of the ions that deliver the pure salt solution. If, for example, a large amount of chloride ions are added to the saturated sodium chloride solution, then the equilibrium deviates in such a way that solid sodium chloride is formed and precipitates (Le Chatelier's principle of "getting rid of the stress"). This way, the position of equilibrium is altered; however, the product of concentrations of sodium ions and chloride ions remain constant: solubility product.

Material: Test tube, pipette; saturated sodium chloride solution, concentrated hydrochloric acid.

Procedure: Fill a test tube to half of its volume with saturated sodium chloride solution and carefully add several drops of hydrochloric acid. Repeat the additions of hydrochloric acid.

Observation: Immediately after the addition of concentrated hydrochloric acid, white crystals of salt are formed and deposited on the bottom of the test tube: sodium chloride.

E6.5 Solubility Equilibrium of Calcium Sulfate

Problem: Many materials like, for instance, sand are completely insoluble and cannot be traced in solution. Many salts dissolve in such minimal amounts that adding them to water will not yield a solution of recognizable amounts of matter. Calcium sulfate powder belongs to this group of hard-to-dissolve salts: it is only possible to prove the existence of small portions of diluted substance simply by evaporating the water of a large portion of the solution or by electric conductivity. Quantitative proof of the dissolved amount is possible through comparing conductivity with similarly concentrated solutions of magnesium sulfate: this salt is very soluble and allows for the preparation of different standard solutions. Because solutions with identical concentrations show the same electrical conductivity, one can ascertain the concentration of saturated calcium sulfate solution by comparing with standard solutions of magnesium sulfate.

Material: Small beakers, equipment for measuring electrical conductivity and tester, balances, graduated cylinder; calcium sulfate, magnesium sulfate.

Procedure: Mix several pinches of calcium sulfate with water; leave the mixture to stand overnight: the solution becomes clear, a big solid residue can be observed at the bottom. Dip the conductivity tester into distilled water, then into the saturated solution. Prepare a 0.1 molar magnesium sulfate solution and dilute it with the help of the graduated cylinder: once at 1:2 and once at 1:10. Measure the conductivity of these solutions and compare with the conductivity of saturated calcium sulfate solution.

Observation: Saturated calcium sulfate solution shows a good electrical conductivity. The saturated solution and the 0.01 molar solution of magnesium sulfate produce similar reading on the conductivity tester. This concentration can therefore be used for the calculation of the solubility product of calcium sulfate.

E6.6 Boudouard Equilibrium

Problem: Reactions involving gases are well suited for direct observation of incomplete reactions through measuring the volume and recognizing the equilibrium. It is also possible through the Avogadro relationship to calculate gas concentrations and apply it to the terms of the law of mass action: equilibrium constants can be demonstrated and calculated this way. The Boudouard equilibrium is part of these equilibria and plays an important role in the furnace process: through the reaction of glowing coal with carbon dioxide, carbon monoxide gas is formed and is ready to reduce iron oxide to iron. In addition, this equilibrium experimentally shows the fact that a mixture of reactants and products exist and that these do not exist separately according to the "left and right side of the equation". Finally, it is shown, that different amounts of solid matter do not influence the equilibrium: if there are 10 g of glowing coal, 100 g or 1000 g – only the presence and the excess amount of coal is important.

Material: 2 syringes, quartz reaction tube, two burners, gas tracing instrument (Draeger tubes for CO), small beakers; carbon pieces, glass wool, carbon dioxide (lecture bottle), limewater.

Procedure: Fill the quartz reaction tube with carbon pieces and close both ends of the tube with glass wool. Attach two syringes, one filled with 50 ml carbon dioxide gas, the other syringe remains empty. Intensely heat the carbon with two burners until glowing; pass the gas slowly over it until the volume remains constant. Attach the gas tracing instrument to the half-gas-filled syringe; suck part of the gas through the testing tube of the instrument. Pass the second part of the gas through a little amount of limewater.

Observation: The highest possible volume of 100 ml is not reached – a constant volume of about 60 ml can be measured. The test tube shows the presence of carbon monoxide, limewater also proves the presence of carbon dioxide.

Tip: Calculations with the aid of the initial and final volume lead to the equilibrium volume of 40 ml CO_2 and 20 ml CO, the comparison of volume proportions with the temperature–volume diagram (see Fig. 6.7) lead to a basic temperature of 650 °C at equilibrium. If one relates these volumes to the entire volume, it is possible to calculate the partial pressures or concentrations – they can be applied to the relationship of the law of mass action in order to calculate the equilibrium constant (see Sect. 6.3).

E6.7 Quantitative Decomposition of Ammonia

Problem: The ammonia equilibrium is an equilibrium that can be thoroughly discussed and evaluated in relation to pressure–temperature changes of gases in association with Le Chatelier's principle (see Fig. 6.10). Unfortunately, it is hardly possible to experimentally produce ammonia from the elements under normal pressure; however, the decomposition of ammonia into the elements is possible. After the decomposition into the elements, the reaction of copper oxide with hydrogen can separate both gases, nitrogen can be proved by the last step. On the other hand, these experiments can lead to the discussion of the Haber–Bosch procedure that is used to manufacture ammonia from nitrogen of the air and from hydrogen – and thereby producing many important nitrogen compounds, i.e. nitrates for artificial fertilizers.

Material: Gas generator, two syringes, two quartz combustion tubes, burner, small gas jar with glass cover, glass bowl; concentrated ammonia solution, sodium hydroxide, pieces of nickel wire, copper oxide (wire form), wood splint.

Procedure: (a) Produce ammonia (Caution: gas is an irritant) by dripping ammonia solution in the gas generator on solid sodium hydroxide. Fill the gas in a syringe, and regulate its volume to 50 ml. The syringe is connected to the other empty syringe above the combustion tube that has been furnished with several pieces of nickel wire. Heat the nickel and pass ammonia across until the gas volume remains constant. (b) Place the second combustion tube which contains copper oxide between both syringes. Pass the gas mixture which resulted from the first reaction over the heated copper oxide until a constant volume is reached. Observe the gas volume and transfer it pneumatically into the small gas jar and test with a burning wood splint.

Observation: (a) From 50 ml gaseous ammonia, 100 ml colorless gas is formed: mixture of hydrogen and nitrogen. (b) By leading the gas mixture over heated black copper oxide, red-brown shiny copper is formed; condensation of water is observed, approximately 25 ml of gas remains. This gas immediately extinguishes the flames of the burning wood splint: nitrogen.

Tip: Students should know that synthesis of ammonia is very important for industries: not only the nitrogen fertilizers are produced through the reaction of nitrogen and hydrogen, also sodium or potassium nitrate for gunpowder, similar explosives and other mixtures.

References

1. Bergquist, W., Heikkinen, H.: Student ideas regarding chemical equilibrium: What written test answers do not reveal. Journal of Chemical Education 67 (1990), 1000
2. Finley, F.N., Stewart, J., Yarroch, W. L.: Teachers' perceptions of important and difficult science content. Science Education, Ausgabe 4, 1982, S. 531–538
3. Tyson, L., Treagust, D.F.: The complexity of teaching and learning chemical equilibrium. Journal of Chemical Education 76 (1999), 554
4. Banerjee, A.C., Power, C.N.: The development of modules for the teaching of chemical equilibrium. International Journal of Science and Education 13 (1991), 358
5. Hackling, M.W. und Garnett, P.J.: Misconceptions of chemical equilibrium. European Journal of Science Education 7 (1985), 205
6. Kousathana, M., Tsaparlis, G.: Students' errors in solving numerical chemical equilibrium problems. CERAPIE 3 (2002), 5
7. Kienast, S.: Schwierigkeiten von Schuelern bei der Anwendung der Gleichgewichtsvorstellung in der Chemie: Eine empirische Untersuchung über Schuelervorstellungen. Aachen 1999 (Shaker)
8. Osthues, T.: Chemisches Gleichgewicht. Empirische Erhebung von Fehlvorstellungen im Chemieunterricht. Staatsexamensarbeit. Muenster 2005
9. Amann, W., u.a.: Elemente Chemie II. Stuttgart 1998 (Klett)
10. Asselborn, W., Jaeckel, M., Risch, K.T.: Chemie heute Sekundarstufe II. Hannover 1998 (Schroedel)
11. Dickerson, R.E., Geis, I.: Chemie – eine lebendige und anschauliche Einfuehrung. Weinheim 1981 (Chemie)
12. Holleman, A.F., Wiberg, E.: Lehrbuch der Anorganischen Chemie. Berlin 1985 (de Gruyter)
13. Dehnert, K., u.a.: Allgemeine Chemie. Braunschweig 2004 (Schroedel)

Further Reading

Banerjee, A.: Misconceptions of students and teachers in chemical equilibrium. International Journal of Science Education 12 (1991), 355

Buell, R.R., Bradley, G.: Piagetian studies in science: Chemical equilibrium understanding from the study of solubility. A preliminary report from secondary school chemistry. Science Education 56 (1972), 23

Cachapuz, A.F.C., Maskill, R.: Using word association in formative classroom tests: Following the learning of Le Chatelier's principle. International Journal of Science Education 11 (1989), 235

Camacho, M., Good, R.: Problem solving and chemical equilibrium: Successful versus unsuccessful performance. Journal of Research in Science Teaching 26 (1989), 251

Gorodetsky, M., Gussarsky, E.: Misconceptualisation of the chemical equilibrium concept as revealed by different evaluation methods. European Journal of Science Education 8 (1986), 427

Griffiths, A.K.: A critical analysis and synthesis of research on students' chemistry misconceptions. In: Schmidt H.J.: Problem Solving and Misconceptions in Chemistry and Physics. ICASE 1994

Gussarsky, E., Gorodetski, M.: On the chemical equilibrium concept: Constrained word associations and conception. Journal of Research in Science Teaching 25 (1988), 319

Gussarsky, E., Gorodetski, M.: On the concept "chemical equilibrium": The associative framework. Journal of Research in Science Teaching 27 (1990), 197

Johnstone, A.H., Macdonald, J.J., Webb, G.: Chemical equilibria and its conceptual difficulties. Education in Chemistry 14 (1977), 169
Kind V.: Chemical concepts: Closed system chemical reactions. Education in Chemistry 29 (2002), 36
Knox, K.: Le Chatelier's principle. Journal of Chemical Education 62 (1985), 863
Maskill, R., Cachapuz, A.F.C.: Learning about the chemistry topic of equilibrium: The use of word association tests to detect developing conceptualizations. International Journal of Science Education 11 (1989), 57
Niaz, M.: Relationship between student performance on conceptual and computational problems of chemical equilibrium. International Journal of Science Education 17 (1995), 343
Niaz, M.: Response to contradiction: Conflict resolution strategies used by students in solving problems of chemical equilibrium. Journal of Science Education and Technology 10 (2001), 205
Pardo, J., Solaz-Portoles, J.J.: Students' and teachers' misapplication of Le Chatelier's Principle: Implications of teaching chemical equilibrium. Journal of Research in Science Teaching 32 (1995), 939
Pedrosa, M.A., Dias, M.H.: Chemistry textbook approaches to chemical equilibrium and student alternative conceptions. CERAPIE 1 (2000), 227
Quilez-Pardo, J., Solaz-Portoles, J.J.: Students' and teachers' misapplication of Le Chatelier's principle: Implication for the teaching of chemical equilibrium. Journal of Research in Science Teaching 32 (1995), 937
Solaz, J.J., Quilez, J.: Changes of extent of reaction in open chemical equilibria. CERAPIE 2 (2001), 303
Van Driel, J.H., De Vos, W., Verloop, N.: Relating students' reasoning to the history of science: The case of chemical equilibrium. Research in Science Education 28 (1998), 187
Voska, K.W., Heikkinen, H.W.: Identification and analysis of student conceptions used to solve chemical equilibrium problems. Journal of Research in Science Teaching 37 (2000), 160
Wheeler, A.E., Kass, H.: Student misconceptions in chemical equilibrium. Science Education 62 (1978), 223
Wilson, J.M.: Network representations of knowledge about chemical equilibrium. Journal of Research in Science Teaching 31 (1994), 1133

Fig. 7.1 Concept cartoon concerning hydrochloric acid

Chapter 7
Acid–Base Reactions

The donor–acceptor principle is an important basic concept in modern chemical education: acid–base reactions, redox reactions and complex reactions explain a huge number of chemical changes. One important group of donor–acceptor reactions are the acid–base reactions: protons (H^+ ions) transfer from one species to another species. One example, in the neutralization of sulfuric acid with sodium hydroxide: a proton is moving from one hydronium ion H_3O^+(aq) of the acid solution to one hydroxide ion OH^-(aq) ion of the hydroxide solution. Broensted's key concept will be considered in this chapter.

In Chap. 8, redox reactions and the transfer of electrons will be discussed and in Chap. 9, complex reactions and the transfer of ligands will be presented. For all these essential concepts, questions will be raised related to students' misconceptions. Finally strategies for teaching and learning to prevent or cure these preconceptions and school-made misconceptions will be discussed.

7.1 Acid–Base Reactions and the Proton Transfer

The term, acid, was at first used by Boyle in the 17th century. It was based on the following phenomena: acids are materials that change the color of certain plant extracts and that dissolve limestone [1]. Glauber, Lavoisier, Davy, Liebig and Arrhenius further developed the acid–base concepts, which were based on the knowledge of the time and which were dominated by information about certain phenomena or substances (see Fig. 7.2).

In 1923, Broensted was the first to develop an acid–base concept that was no longer related to substances, but rather to the function of particles. Acids are proton donors and are capable, with suitable reaction partners, to donate protons to base particles or proton acceptors, i.e. protolysis or proton transfer reaction. For example, HCl molecules, as acid particles, transfer protons when colliding with H_2O molecules (see Fig. 7.3). In this sense, the proton donors of pure sulfuric acid are H_2SO_4 molecules, the acid particles of the sulfuric acid solution are the hydronium ions (or also the hydrogen sulfate ions in concentrated solutions). In weak acids, the protolysis equilibrium is to be considered, equilibria and their constants are well defined.

Fig. 7.2 Historical acid–base concepts [2]

From Arrhenius' theory, almost identically relevant dissociation equilibria and corresponding dissociation constants can be written. When discussed in class they may cause language problems for the learners: **from Arrhenius' point of view, acids are substances but, from the view of Broensted, acids are small particles.** In this regard, it is no wonder that considerable misconceptions occur

> Dissolving hydrogen chloride gas in water is a chemical reaction. A Cl⁻ ion forms during the collision of one HCl molecule with one H₂O molecule through heterolytic separation of the H–Cl bond. The simultaneously transferred H⁺ ion (a proton) joins a free electron pair of a H₂O molecule:
>
>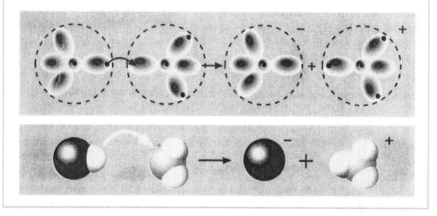
>
> No free H⁺ ions are formed (as was postulated by Arrhenius in 1883); the proton which is initially bonded to the chlorine atom through one electron pair separates from these electrons. Then it "slips" into one of the free electron clouds of the oxygen atom which is connected with two other protons.

Fig. 7.3 Mental model of proton transfer from one HCl molecule to an H₂O molecule [3]

when acid–base concepts are discussed in chemistry classes. These and other misconceptions will be addressed below.

7.2 Misconceptions

Some misconceptions related to acids and bases are found in the chemical education literature. Examples of more misconceptions appear in the research done by Sebastian Musli [4]. He developed a questionnaire and gave it to about 100 students at the Secondary Level II of German academic high schools [4]. Unusual and interesting statements from students have been quoted relating to acids, specifically on the differences between pure acids and acidic solutions, on neutralization, and on differences between strong and weak acids.

Acid concepts. Astonishingly, only acids are attributed an "aggressive effect", although bases also have this attribute: "Acids eat away, acids destroy, and acetic acid is a destructive and dangerous substance in chemistry, not used in normal everyday life" [4]. Similar statements are also quoted in other sources:

"An acid is something which eats material away or which can burn you; testing for acids can only be done by trying to eat something away, the difference between a strong and a weak acid is that strong acids eat material away faster than weak acids" [5]. Barker comments on these students' statements as follows: "No particle ideas are used here; the students give descriptive statements emphasizing a continuous, non-particle model for acids and bases, some including active, anthropomorphic ideas such as 'eating away' (...)" [5].

Regarding the question, "what do you understand by the term acid or base?", many students respond with a pH value ("acids have a small pH value") or with simple, but misconceived phenomena, which are formulated as follows: "Acids eat away, are dangerous, yellow, red or acidic" [4]. Other statements describe acid concepts, which have been mainly learned and remembered: Approximately 15% of the answers show the Arrhenius concept (acids contain H^+ ions); approximately 30% show the Broensted concept (acids release protons), whereby it is not certain if students correctly understand the notion of acids as acid particles.

In the additional exercise, "give examples for atoms/ions/molecules that are acids or bases", mostly formulas for hydrochloric acid, sulfuric acid and acetic acid are noted and for bases, mainly "NaOH". Regarding the Broensted concept, the correct answers for base particles, i.e. the hydroxide ions have only been listed in about 15% of the cases, at the same level as hydronium ions as particles in diluted acidic solutions.

Sumfleth [6] also states, that the idea of proton transfer may be learned by students but cannot be applied in a new context. Sumfleth and Geisler [7] show that students accept the Broensted definition, but bases are interpreted mostly based on the Arrhenius' idea. Therefore, the knowledge about Broensteds' concept cannot be transferred to new contexts. Sumfleth states that most "students cannot really apply acid–base theories, especially at the advanced levels. This is also evident for students who have chosen chemistry as their major".

Sumfleth discovered several misconceptions when evaluating special questionnaires and student concept maps [8]. After learning the Broensted concept of acids and bases in advanced lessons, students have a lot of difficulties with the idea of an acid. They tend to think in three directions:

1. acids as pure substances like the gas hydrogen chloride (HCl),
2. acids as solutions like hydrochloric acid, containing H^+(aq) ions and Cl^-(aq) ions,
3. acids as particles like hydronium ions, H_3O^+(aq).

Mostly, students mix up all three ideas. They speak of substances: "hydrochloric acid gives protons"; they think protons come out of the nucleus of atoms or ions: "the other particle should be radioactive", etc. Students have problems switching from the level of substances to the level of particles and they like – even in advanced classes – to stay on the level of substances: "hydrogen chloride plus acid gives hydrochloric acid". When discussing corresponding acid–base pairs, students do not deal appropriately with the level of particles, they prefer

7.2 Misconceptions

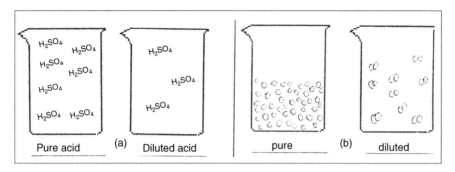

Fig. 7.4 Examples of misconceptions on sulfuric acid solution [4]

to state: "hydrogen chloride and water form the corresponding acid–base pair". Schmidt [9] describes similar findings.

Pure Acids and Acidic Solutions. In this exercise, the students are supposed to state the similarities and differences between pure sulfuric acid and the 0.1 molar solution, and to schematically draw the smallest particles in two model beakers (see Fig. 7.4). Correct answers regarding the hydronium ions and sulfate ions in dilute solutions can be found in only 10% of the answers or model drawings. Approximately 45% of the answers approach it from the dilution effect: either the drawings depict for example, symbols for sulfuric acid molecules with larger distances in the solution or hard to understand spherical models (see Fig. 7.4).

Many other answers offer different claims: "pH value of pure acid is less; pH values are different for acids and acidic solutions (without mentioning pH value or differences), the densities vary; pure acids are much more corrosive, are more amenable to reactions than the solution, etc.".

Only about 10% of the students gave the correct verbal answers *and* included appropriate model drawings with the expected ion symbols for the dilute solution. The surprising thing is that two students who gave a correct verbal answer regarding the "dissociation in dilute sulfuric acid solution" did not note any ion symbols. The first student used the H_2SO_4 symbol for the acid solution (see (a) in Fig. 7.4); approaching it from the particle model of matter the second student offered hard to interpret close spherical models (see (b) in Fig. 7.4). The term "dissociation" appears, however, to be totally misunderstood by these two students.

Thirty percent of those questioned neither gave a verbal answer nor produced a model drawing. This shows that a significant percentage of Secondary Level II students of German academic high schools do not understand the subject of acid–base reactions, although these reactions are a prerequisite for concepts such as differences between strong and weak acids, equilibria and acid constants of weak acids and bases.

pH Value. Several students elucidated on the pH value and pH scale: they correctly wrote the "negative logarithm of H^+ ion concentration", without showing where and how they can apply these definitions. Many students have uselessly changed these definitions into "the pH value is the negative logarithm to the base ten". There are even purely phenominalistic answers: "pH 7 is neutral to the skin; pH can be found in shampoos; it has some connection to skin compatibility".

The second part of the exercise challenges the students to come up with solutions which have a pH of 1; only 15% mentioned a 0.1 molar solution of hydrochloric acid; 30% reported acid solutions like hydrochloric acid and sulfuric acid without mentioning the concentration; and 45% did not even answer the question. This distribution of answers confirms a lack of true understanding of pH values by students of the Secondary Level II – or an inability to apply and to understand this subject matter completely.

Neutralization. In this exercise, it was stated that "hydrochloric acid reacts with sodium hydroxide solution". The students were asked first to show chemical equations using the types of involved particles. Approximately 80% of the students were able to write the common equation: "HCl + NaOH → NaCl + H$_2$O".

Half of the students noted the reaction equation with ion symbols and expressed that the H$^+$(aq) ions and the OH$^-$(aq) ions react to produce H$_2$O molecules. Most of the students stated, that "NaCl" is formed without showing the correlative sodium ions and chloride ions; some even offer "solid NaCl" or "NaCl crystals" as reaction products.

Similar ways of thinking are established in the second part of the question about reacting particles: besides the approximate 40% of correct answers, many other answers show the opinion that "all particles" or "positive and negative charges" react with each other, that "NaCl molecules are formed during the reaction". Many students are very insecure with the formulation of the reaction equation (see in Fig. 7.5). Also the question about the name of this reaction is

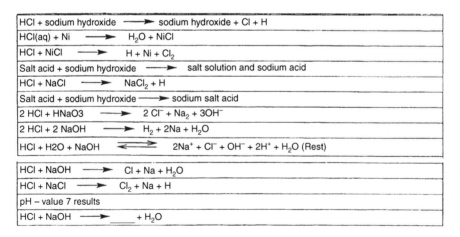

Fig. 7.5 Examples of incorrect reaction symbols on acid–base neutralization [4]

7.2 Misconceptions

correctly given as "neutralization" by only 35% of the students; other students prefer a redox reaction, a chlorination or titration.

Barke [8] also found, through his empirical studies on the concept of neutralization, that in the first step students conceive it as obtaining sodium chloride solution. In the second step the evaporation of water from salt solution and the formation of solid sodium chloride should be interpreted. Many students thought that sodium ions and chloride ions were linked together to form NaCl molecules, that the charges are thereby "neutralized" (see Fig. 7.6). The students mix concepts of acid–base neutrality and electrical neutrality: "NaCl solution is neutral, because Na^+ ions and Cl^- ions bond and charges are neutralized; the charges are neutralized in crystals; equal numbers of Na^+ ions and Cl^- ions neutralize each other; crystalline material is the result of a neutralization, therefore it is neutral; the charges balance each other out in the compound, the compound is neutral" [8]. It is therefore recommended to designate the term "neutrality" entirely to the acid–base topic and not to argue with electrical neutrality of ions in solutions or in crystals, but rather to argue on the basis of "equilibrium of anions and cations" or with the "equilibrium of charges" [8].

Concerning the idea of neutralization, Sumfleth [9] found that students think along the lines of acid–base equilibria: "After neutralization, sodium chloride solution contains the same amount of hydrochloric acid and sodium hydroxide solution; with neutralization there exists equilibrium of acid and base". The reaction of solid sodium chloride with water to form a salt solution is correctly recognized as dissociation by only 15% of the students; more than 35% assume an inversion of neutralization: "After the reaction of sodium chloride with water, the same amounts of acid and base are found in the solution". Considering the reaction of solid sodium phosphate and water and observing a basic solution, students do not hesitate to explain: "OH^-(aq) ions and $Na_3PO_4H^+$(aq) ions are produced by the reaction". In these examples, one can see that the term "dissociation" of salt in water seems to be very misunderstood [9].

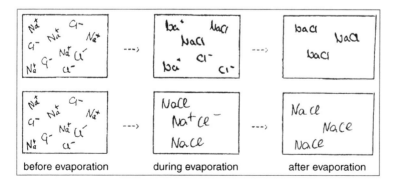

Fig. 7.6 Misconceptions regarding evaporation of water from a salt solution [8]

Schmidt [10] asked about the reaction of sodium acetate with water, he was hoping to hear that $OH^-(aq)$ ions form the basic solution. However, most students stated that the pH of 7 results and made comments like, "sodium acetate molecules dissociate in water to form ions. These ions are capable of reacting with the ions from the water. As the produced NaOH and CH_3COOH molecules dissociate instantaneously, the result is an equilibrium with the concentration of OH^- and H_3O^+ ions remaining unchanged" [10]. "By dissolving salts in water the ions only get hydrated, H_2O molecules surround the ions of the salt, OH^- and H_3O^+ ions are not involved; if you mix an acidic solution with a basic solution, a salt will be produced which contains neither OH^- ions nor H_3O^+ ions" [11].

Schmidt also asked about the neutralization of acetic acid and sodium hydroxide solution and found that most students came up with a pH value of 7. These students stated that NaOH and CH_3COOH are neutralizing, a salt is formed, also water; some water molecules dissociate to produce H_3O^+ ions and OH^- ions [11]. The word "neutralization", in reference to the reaction of weak acids, confuses students totally. Schmidt cited some comments like, "you cannot call this neutralization just because we have a slightly basic pH value; the product is slightly basic, so why call it neutralization? Neutralization does not mean that the final pH value is neutral; please define it a little more precisely: does it have a neutral pH value or acid and base being balanced? Well, I could neutralize only partly and thus it would be right to call this neutralization; it is not good to call this a neutralization, we had two solutions and made a relatively neutral one out of them, but in the end we have a pH value that is not neutral; neutralization is not like either dead or not dead, it happens gradually, it is the path towards neutrality" [12].

Strong and Weak Acids. Sumfleth [6], Sumfleth and Geisler [7] describe the common misconception, that for most students acid strength is solely based on the pH value of solutions. Thus, it is possible for them to determine the acid strength in an experiment by using acid–base indicators, e.g. universal indicator that can quantitatively estimate the pH value through color comparisons. Students overlook the fact that by taking a 1 M hydrochloric acid solution with a pH value of 0, one can dilute every larger pH value up to almost 7. The questions regarding acid strength as concentrations of acid molecules and ions and mixing those ideas, causes confusion to students, the results are incorrect concepts.

In a related question of the questionnaire, students were asked to compare and contrast 0.1 M solutions of hydrochloric acid (HCl) and acetic acid (HAc); and in addition students were requested to draw schematic model drawings of possible atoms/ions/molecules [4]. Approximately half of the students gave no answers concerning similarities and differences, 20% mentioned the acid strength, and 10% noted the pH value as differences. A higher pH value was however mistakenly attributed to hydrochloric acid. Mostly acetic acid was regarded as "the stronger acid because a larger I-effect of the methyl group can be registered at CH_3COOH molecules and therefore the proton can more

7.2 Misconceptions

easily split off" [4]. This quotation shows that the treatment, which coincidentally took place in the half year of the studies in organic chemistry, lead the students to associations on arbitrary contents, which they did not properly understand.

Only to 15% of the student showed appropriate acetic acid molecules and the related ions in their model drawings (see first drawing in Fig. 7.7). To the same degree, students draw correct ion symbols but no molecule symbols, or they merely imagine only molecules and no ions (see Fig. 7.7). From this data one can easily conclude that these students have not understood the differences between strong and weak acids.

Several further drawings show other incorrect mental models: only the related ions or only the molecules (see Fig. 7.8). Structural symbols for sulfuric acid molecules or acetic acid molecules are, surprisingly enough, correct. The fact that several students draw incomprehensible sphere pattern in the model beakers, shows that the expected model drawings are not common in their lessons. Almost 50% of students avoid any drawing whatsoever – they are not capable of conceiving a mental model or of putting their ideas in such model drawings.

In another exercise, the students were supposed to answer the question regarding differences between strong and weak acids; they were requested to mark one of the following alternatives: "molar mass of acid molecules; concentration of hydronium ions (pH value); level of dissociation (K_s value); strength of hydrogen bonding; density of acid; concentration of metal ions". They were also requested to explain their choice.

The student scores are almost completely concentrated on the incorrect answer "pH value" (40%) and the correct answer "level of dissociation" (55%).

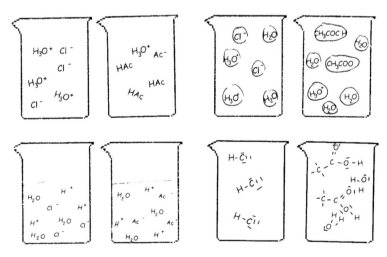

Fig. 7.7 Examples for appropriate and inappropriate mental models on weak acids [4]

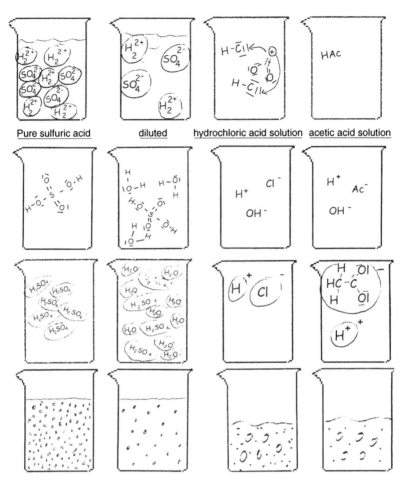

Fig. 7.8 Other examples of incorrect model drawings on acids and acidic solutions [4]

Luckily, in the reasons given for "level of dissociation", the arguments are based on equilibrium and on the differences of dissociation – unfortunately, only a small amount of the students provided an explanation for the correctly marked answer. It can therefore be assumed, that almost half of the students see the pH value as a criteria for differentiating between strong and weak acids and they therefore confirm the expected misconceptions, also found by Sumfleth [6].

Summary. One can sketch the following aspects of misconceptions: Many students assign the characteristics of acids as "being destructive or corrosive". The hydroxides and concentrated basic solutions are not mentioned in this aspect.

Students often do not differentiate between pure acids, which, except for hydrochloric acid are made up of molecules; and diluted acid solutions, which are mainly made up of ions. Concerning the term acid, mental models are mixed between Arrhenius' statement "acids are substances which contain H^+ (aq) ions" and Broensteds' concept of "acids are particles which release protons". The following misconception often shows the mixing of both concepts: "hydrochloric acid is a proton donor".

Regarding neutralization, one tends to use the overall equation, and the salt formation is pushed to the front. In many cases, neutralization is transferred to the discharging of ions in the crystallization of a salt crystal: "ions join to form salt molecules".

Differences between strong and weak acids lead – as expected – to incorrect statements about the concentration of the solutions, and correspondingly to the pH value. Dissociation or protolysis level have surely been dealt with in the lessons, but probably only talked about and therefore never became established in student's minds.

Mental models regarding the small particles of acids and bases seem altogether to be grounded on putting together misunderstood formulas and reaction symbols. Mental models have not been looked at in the form of model drawings and therefore are only tentatively present or are misunderstood. Such model drawings should play a central role in proposed lessons that may prevent or cure misconceptions.

7.3 Teaching and Learning Suggestions

Phenomena. In the first step, one could clearly look at **aggressive effects** – the well-known phenomena of acidic *and* basic solutions. Concerning acids, one can demonstrate the spectacular reaction of concentrated sulfuric acid with sugar (see E7.1) or that of the behavior of acidic solutions with metals (see E7.2). One should discuss, in both cases, the statement that "an acid is something which eats material away" [5], and can demonstrate that other substances are produced by those acid reactions: sulfuric acid and sugar produce black carbon and steam; metals react to produce hydrogen and a salt solution, from which solid white salts may be obtained by the evaporation of water. In addition, acidic household cleaners like those that remove lime deposits could be introduced: one could demonstrates that when lime deposit is removed by a cleaner solution, salt solution and carbon dioxide gas are produced (see E7.3).

Regarding the aggressiveness of metal hydroxides, one could also mention "drain cleaners" as a common household aid (see E7.4): organic materials such as cloth fabrics, paper or hair dissolve when these concentrated hydroxide solutions are used. In addition, a diluted hydroxide solution has aggressive effects on clothing: as soon as the water evaporates, the aggressive effect of the hydroxide sets in and holes appear in the clothing. In this context, one should always clearly differentiate between solid sodium hydroxide and its basic solution in water.

The same phenomenon may be discussed for sulfuric acid: spills of diluted sulfuric acid on clothing produce holes because the evaporation of water causes the acid to become more concentrated which has an aggressive effect on clothing. It is different in the case of hydrochloric acid or acetic acid: in both solutions, the acidic substances vaporize first and finally, the water evaporates.

In the case of acids, students should also be aware that the reaction of diluted solutions can be beneficial: stomach acid is diluted hydrochloric acid which is essential for the digestive process, citric acid or acetic acid are substances which are useful in the kitchen, phosphoric acid is an acidic ingredient of cola soft drinks, and very diluted sulfuric acid can often be found in mineral water.

Other phenomena which shouldn't be ignored are reactions of **acid–base indicators,** which change color depending on the concentration of acidic or alkaline substances. All students should be made aware of these color changes (see E7.5). In this regard, the universal indicator is an especially excellent example: it does not merely show a neutral reaction of solutions through the green color, but also has different shades of red and shows the pH value of acidic solutions. It also detects basic solutions with various shades of blue. Therefore, this indicator provides an excellent method of demonstrating the terms acidic, neutral and basic.

For introducing the idea of pH values, it is possible at this point, to use the universal indicator to show pH values by different colors. In addition a pH meter may be demonstrated, the pH values interpreted and compared with those of the indicator paper – no hydronium concentrations or logarithms are necessary at this stage. Using these methods, even young students at elementary schools may test several household solutions (see E7.6), and may demonstrate the results on a chart (see Fig. 7.9).

Acid–Base Concepts. In the lesson, after becoming familiar with many phenomena on acids and bases as substances, the question is raised as to whether the substance-related Arrhenius concept should be taught, or the particle-related Broensted concept – or the genetic development of both concepts in the form of historically oriented lessons (see Fig. 7.2). Tests on electrical conductivity of solutions of strong acids and bases (see E7.7) confirms that acidic solutions contain H^+(aq) ions and basic solutions contain OH^-(aq) ions. After stating the existence of these ions, one can discuss model drawings and emphasize that the (aq) symbol denotes the complete separation of the ions by hydration (see Figs. 7.10 and 7.11).

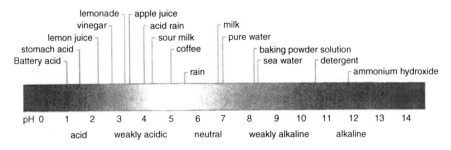

Fig. 7.9 pH values of several common and environmental chemicals [2]

7.3 Teaching and Learning Suggestions

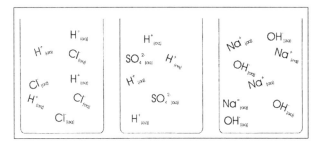

Fig. 7.10 Mental models of acidic and basic solutions

Next, these and similar drawings should be attempted by the students themselves. For all concentrations of hydrochloric acid, mental models of ions are correct. This is not the case when one compares this with pure sulfuric acid or pure nitric acid: a pure acid contains molecules and should be described by molecular symbols.

Mental models are thus produced of both acid concepts: those of Arrhenius and of Broensted. The Arrhenius concept explains some phenomena in the area of acids and bases, for instance, the neutralization of solutions of strong acids and bases. Terms like weak acids or derived concepts like acid constants or buffer already reach the limits of Arrhenius' concept.

Broensted Concept. This concept is much broader – but acid *particles* and base *particles* are considered opposite to all other concepts. It is therefore worth basing the lessons as soon as possible on acids as donor *particles* and on bases as acceptor *particles,* and to consistently using them in the subject terminology and in model drawings.

One possibility of introducing the concept is the production of gaseous hydrogen chloride from sulfuric acid and common table salt (see E7.8) and subsequent reaction of the produced gas with water to form hydrochloric acid (see E7.9). In the first part, H_2SO_4 molecules donate protons to Cl^- ions from sodium chloride crystals to form HCl molecules and HSO_4^- ions:

$$H_2SO_4 \text{ molecules} + Cl^- \text{ ions} \rightarrow HCl \text{ molecules} + HSO_4^- \text{ ions}$$

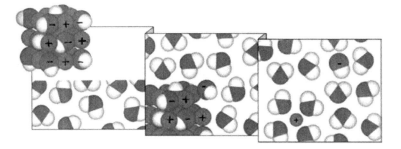

Fig. 7.11 Mental model for reactions of sodium hydroxide with water [14]

To visualize this reaction, a molecular model set should be used to build the structure of the involved species and to simulate the proton transfers (see Fig. 7.12).

In the second part, the HCl molecules are donating protons to H_2O molecules to produce H_3O^+(aq) ions and Cl^-(aq) ions, molecular models should visualize it (see Fig. 7.12):

$$\text{HCl molecules} + H_2O \text{ molecules} \rightarrow H_3O^+(aq) \text{ ions} + Cl^-(aq) \text{ ions}$$

Referring back to the first part, the H_2SO_4 molecules are the proton donors or acid particles; in the second part the HCl molecules are donating protons. In the resulting hydrochloric acid the H_3O^+(aq) ions are now the acid particles: one has to prevent the obvious misconception that there are HCl molecules in hydrochloric acid. If these reaction symbols and terminology were repeated in other examples, then the students would gain a better understanding of the concept. By diluting pure sulfuric acid, the usual dilution effect does not explain the strong exothermic reaction (see E7.10) – new particles occur by the transfer of protons, the transfer should be visualized by molecular models (see Fig. 7.12):

$$H_2SO_4 \text{ molecules} + 2H_2O \text{ molecules} \rightarrow 2H_3O^+(aq) \text{ ions} + SO_4^{2-}(aq) \text{ ions}$$

It depends on the amount of water: a relatively concentrated sulfuric acid solution also contains HSO_4^- ions, which are acid particles and could donate protons. In such a concentrated solution, we have two kinds of proton donors: H_3O^+(aq) ions and HSO_4^-(aq) ions!

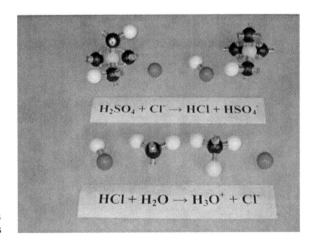

Fig. 7.12 Molecular models to visualize proton transfers

7.3 Teaching and Learning Suggestions

Another example shows the exothermic reaction of calcium oxide with water (see E7.11), this time the O^{2-} ions are the base particles, whereas H_2O molecules function as acid particles:

$$O^{2-} \text{ ion} + H_2O \text{ molecule} \rightarrow OH^-(aq) \text{ ion} + OH^-(aq) \text{ ion}$$

Because H_2O molecules can take on both the function of an acid as well as the function of a base, they are called amphotheric particles. Another amphotheric particle is the HSO_4^- ion: depending on the reaction partner, it reacts as an acid particle or as a base particle.

In solid hydroxide, the OH^- ions already exist as basic particles in the crystal lattice. If solid sodium hydroxide reacts with water in a strong exothermic reaction, the already existing Na^+ ions and OH^- ions are hydrated and in this way separated from each other (see Fig. 7.11):

$$Na^+OH^-(s) \rightarrow aq \rightarrow Na^+(aq) \text{ ions} + OH^-(aq) \text{ ions}$$

pH Value. An operational use of the pH term should occure naturally after students have tested various household solutions with universal indicator paper with pH scale – they don't have to deal with any logarithm (see Fig. 7.9). It is more difficult to teach the quantitative meaning of pH value: one has to work with concentrations, the logarithm and the mol term. In this case, it is advantageous to relate the meaning of 1 mol to a specific amount of small particles and to decide the type of particles, which are to be counted: 18 g water do not contain "1 mol of water", but rather "1 mol of H_2O molecules". A liter of 1 M hydrochloric acid contains 1 mol of $H_3O^+(aq)$ ions and 1 mol of $Cl^-(aq)$ ions; the concentration is equal to 1 mol/l for both kinds of ions. Dilution in the volume ratio 1:10 results in a solution with the $H^+(aq)$ ion concentration of 0.1 mol/l, the dilution 1:100 leads to the 0.01 M or 10^{-2} M solution.

Because the pH values ("*potenta hydrogenii*" or potential of H_3O^+ ions) depend on the concentration of the hydronium ions, it is necessary to go into this in more depth. If one writes the concentration in the form of decimal numbers, the pH can easily be expressed as an exponent with negative sign: a 10^{-2} molar hydrochloric acid has the pH value of 2; the 10^{-3} molar solution the pH value of 3.

It is not necessary to speak of "negative logarithms to the base ten". It is more important to understand that the dilution of an acid by 1:10 results in an increase of the pH value by 1 (see E7.12). That also means, that the pH value 4 of mineral water correlates to a concentration of $H_3O^+(aq)$ ions of 10^{-4} mol/l: the mineral water therefore shows a minimal acidic effect. If one takes the 10^{-1} molar hydrochloric acid solution in the laboratory and wants to have the same pH 4 solution, one has to dilute this hydrochloric acid to the factor of 1:1000 (see E7.12). Should the pH value 6 of rain water be simulated (rain water is slightly acidic through the reaction of water with carbon dioxide in the air), then 1 ml 10^{-1} molar hydrochloric acid has to be further diluted with distilled water to 100 l (volume of a bathtub).

Through these experiments, one gets a first impression of ideas related to pH values. If one teaches the combination of the ion product of water with the concentration of hydronium ions and hydroxide ions in aqueous solutions, then it is possible to derive the pH values of alkaline solutions. Tables and graphs are extremely useful for such considerations (see Fig. 7.13).

If one wants to show different pH values arising not from an acidic solution but from an alkaline solution, a very popular experiment can be done. Saturated calcium hydroxide solution ("limewater") is tested using pH meter and conductivity tester. The bubbling of carbon dioxide gas through the solution causes the well-known white precipitation; after bubbling for a further two minutes that precipitation dissolves again (see E7.13). pH values and electrical conductivity are measured during all reactions; the results can be compared, discussed and explained.

Neutralization. In the interpretation of the reaction of acidic solutions with alkaline solutions (see E7.14), the overall equations of the type "HCl + NaOH → NaCl + H$_2$O" are mainly used in most lessons, thereby creating the discussed misconceptions. Generally, learners do not recognize the corresponding ions in either the reacting solutions or the formed sodium chloride solution. In addition, students do not really differentiate between "salt" or "salt solution", mostly they develop mental models in the direction of "salt molecules".

If ion symbols in the reacting solutions have been successfully drawn (see Fig. 7.10) and neutralization reaction is then discussed based on these drawings, one automatically comes to the correct interpretation that those hydronium ions, through the transfer of protons to hydroxide ions, react with each other to form water molecules. During the reactions, the hydrated sodium ions Na$^+$(aq) and hydrated chloride ions Cl$^-$(aq) are not involved (see Fig. 7.14). These ions are known as "spectator ions": they exist as "spectators"

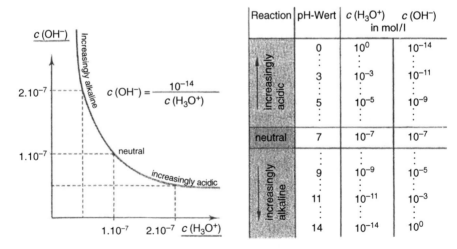

Fig. 7.13 Correlations between pH values and concentrations [13, 14]

7.3 Teaching and Learning Suggestions

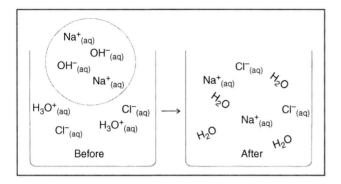

Fig. 7.14 Mental model on neutralization of hydrochloric acid by sodium hydroxide solution

and do not participate in the reaction; they form the resulting sodium chloride solution. Solid salt would appear only after the evaporation of the water from the produced salt solution (see E7.14) – in that case, it is necessary to describe the ionic lattice in the salt crystals produced (see also Chaps. 4 and 5).

Neutralization reactions can also be looked at through conductivity tests. If a conductivity titration is performed with diluted hydrochloric acid and diluted sodium hydroxide solution (see E7.14), one would attain a certain minimum of electrical conductivity at the equivalence point, but it is not zero (see left diagram of Fig. 7.15): Na^+(aq) ions and Cl^-(aq) ions are left behind in the solution and are responsible for the minimal conductivity. The decrease in conductivity is often explained by the misconception that the absolute number of ions decreases and that by the usual ionic equation "from an initial number of four ions, only two remain":

$$H_3O^+(aq) + Cl^-(aq) + Na^+(aq) + OH^-(aq) \rightarrow Na^+(aq) + Cl^-(aq) + 2H_2O(l)$$

It is better to imagine that in the essential step water molecules result and do not therefore contribute to the electrical conductivity:

$$H_3O^+(aq) + OH^-(aq) \rightarrow 2H_2O(l)$$

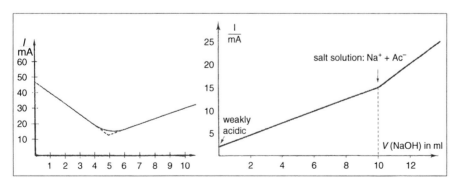

Fig. 7.15 Diagrams of conductivity titrations of hydrochloric acid and acetic acid [15]

One should be aware, that H_3O^+(aq) ions react to H_2O molecules, that however each required OH^-(aq) ion brings in a Na^+(aq) ion: formally, each H_3O^+(aq) ion in the acid solution is replaced by a Na^+(aq) ion from the alkaline solution (see Fig. 7.14). The lowering of the conductivity has to do with the different mobility of ions: the charges of H_3O^+(aq) ions and OH^-(aq) ions move faster in their solutions and contribute more to the conductivity than Na^+(aq) ions or Cl^-(aq) ions. In this way, by obtaining the sodium chloride solution, the minimum of electrical conductivity results (see left in Fig. 7.15).

The conductivity can go down to zero if an insoluble salt is produced by the neutralization reaction. If one carries out a conductivity titration of barium hydroxide solution with diluted sulfuric acid, it is possible to observe a decrease in electrical conductivity (see E7.15). However, because additional white barium sulfate is precipitating, conductivity goes down to almost zero: Ba^{2+} ions and SO_4^{2-} ions do not remain mobile, but form the ion lattice of barium sulfate crystals.

During neutralization of strong acids and bases only H_3O^+(aq) ions react with OH^-(aq) ions. If equivalent portions of acidic solution and basic solution of the same temperature are added and the maximum of increasing temperatures is measured, one observes an increase in temperature which is always the same (see E7.16). These measurements prove that only water molecules are formed in these reactions, the other ions do not change: spectator ions! Calculating the energy change, one finds the constant of heat of neutralization of 56 kJ/mol.

In each case, it is necessary to discuss the scientific mental model on neutralization of strong acids and strong bases with students and to sketch related model drawings (see Fig. 7.14): appropriate interpretations are better retained through these model drawings, and sustainable mental models will be developed in the cognitive structure of students!

Weak Acids and Bases. The term "weak" suggests in itself the following most common misconception: weak acids are "weakly concentrated". It may well be that during students' lessons, protolysis equilibrium of acetic acid was used as an example, maybe perhaps even equilibrium constants came into play, and pH values of specific acetic acid solutions were measured or calculated – however, only a few students are able to comprehend and connect all these facts to develop the scientific idea about weak acids.

In order to look at the degree of protolysis, it is advisable to use convincing experiments. If the pH values of 1.0 molar and 0.1 molar solutions of two acids,

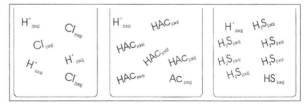

Fig. 7.16 Mental models of two weak acids compared with strong hydrochloric acid

hydrochloric acid and acetic acid, are measured with a calibrated pH meter (see E7.17), one gets the expected pH values of 0 and 1 for hydrochloric acid solutions – but not for the acetic acid solutions: approximately pH values of 2.4 and 2.9 are measured. When this happens, a classic cognitive conflict arises: "what is so different about acetic acid"?

If the 0.1 molar acetic acid solution shows a pH of nearly 3, the concentration of H^+(aq) ions should be 10^{-3} mol/l. Because the concentration of the acetic acid molecules ("HAc molecules") starts with 10^{-1} mol/l, only 1% of the HAc molecules protolyse into ions. In a model drawing, one should draw 99 HAc molecules compared to only 1 H_3O^+(aq) ion and 1 Ac^-(aq) ion – in every case the number of molecules must be higher than the number of ions (see Fig. 7.16).

Another comparative experiment can also convince students. A syringe is filled with 24 ml of hydrogen chloride gas, a second syringe with 24 ml of hydrogen sulfide gas (see E7.18): at room temperature, they each contain 1 mmol of molecules, i.e. the same amount of approximately 6×10^{20} molecules. 10 ml each of distilled water are added and the syringe is intensively shaken: both gas portions will completely dissolve. If 1 mmol molecules have dissolved in 10 ml of solution, a concentration of 0.1 mol/l or 10^{-1} mol/l results in both cases. The pH values of both solutions are then measured: the expected pH value 1 is observed by hydrochloric acid; the pH value of hydrogen sulfide solution, in comparison, is not 1 but 4! If one calculates the protolysis level of hydronium ions and hydrogen sulfide ions, then protolysis level is 0.1%: only 1 of 1000 H_2S molecules forms the related ions (see Fig. 7.16).

Appropriate model drawings should be presented and studied (see Fig. 7.16), the idea of a static protolysis level introduced. Later the dynamic concept can be added to the static concept, and can clarify that a dynamic equilibrium exists (see also Chap. 6).

It becomes clear that both acetic acid solutions and hydrogen sulfide solutions cannot reach the small pH value of 0 or 1 as in hydrochloric or sulfuric acid solutions (see Fig. 7.16). An acid is regarded as weak because of its low protolysis level. Acids like hydrochloric acid, nitric acid or sulfuric acid are strong acids because nearly a 100% protolysis exists – any desired concentration can be attained by dilution (e.g. pH values of 0, 1, 2 or 3, etc.).

In addition, many metal sulfides can be precipitated using H_2S solution. An example of such a precipitation product is solid black copper sulfide, which is deposited by the reaction of copper sulfate solution with H_2S solution (see E7.18 "Tip"). At the same time, it is observed that the H_2S solution changes from pH 4 to pH 1, to an acidic sulfuric acid solution. The reason is the increase of H_3O^+(aq) ions through the extraction of S^{2-}(aq) ions from the equilibrium:

$$H_2S(aq) + 2H_2O\ (l) \leftrightarrows 2H_3O^+(aq) + S^{2-}(aq)$$

The pH dependence of protolysis equilibria on metal sulfide deposits is the basis of the well-known historic separation procedure for many metal cations by precipitation of many different metal sulfides. But for all these considerations the idea of the dynamic protolysis equilibrium is necessary.

Additionally, electrical conductivity measurement helps in the understanding of protolysis equilibrium for weak acids. The comparison of equally concentrated strong and weak acids supplies the much lesser conductivity for weak acids. If one carries out a conductivity titration with acetic acid solution (see E7.14), one gets very different forms of conductivity curves in comparison to the titration of strong acids (compare both diagrams in Fig. 7.15). Titrating with sodium hydroxide, the measured values do not decrease but they rather increase. In this titration, a very low concentration of hydronium ions reacts with hydroxide ions, but mostly the large number of HAc molecules is transferring protons to OH^-(aq) ions: HAc molecules are replaced by Ac^-(aq) ions and therefore the increase in conductivity is explained. Later, after the equivalent point is reached and an excess of hydroxide ions appears, the curve increases more steeply. As a summary, there are two kinds of neutralization reactions:

$$HAc(aq) + OH^-(aq) \rightarrow H_2O(l) + Ac^-(aq)$$

$$H_3O^+(aq) + OH^-(aq) \rightarrow 2H_2O(l)$$

Strong acids do not remain strong forever. If hydrogen chloride is dissolved in pure acetic acid instead of water, it turns into a weak acid in this solution (see Fig. 7.17). Just as much as it depends on its partner to react as an acid or a base, in a similar manner, it also depends on the solution particles which meet acid molecules in order to function as strong acids or a weak acids.

If these subjects are taught with related ions as existing particles in solutions, if concentrations of molecules and ions are compared and discussed, then appropriate mental models can be constructed. If students also draw correlating model drawings for most experimental experiences (see Figs. 7.10, 7.11, 7.12, 7.14, 7.16 and 7.17) and compare and discuss these and other models, then the comprehension for acid–base reactions can have a long lasting effect, thus the learning lead to sustainable understanding of chemistry!

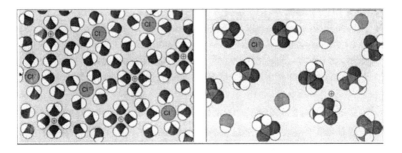

Fig. 7.17 Mental models of HCl solutions in water and in pure acetic acid [2]

7.4 Experiments on Acids and Bases

E7.1 Decomposition of Sugar Using Sulfuric Acid

Problem: Students are familiar with acids and know that they belong to the category of dangerous substances. When they think of acids they often think of destruction: "everything is decomposed or destroyed by acids". On the one hand, it is important to teach the necessary functions of acids, i.e. as preservation methods, as spices or as stomach acid in the digestive system. On the other hand, one should show the aggressive properties by pointing out the reaction products emphasizing more of a chemical process than of the complete destruction of material. The reaction of sugar with concentrated sulfuric acid is a spectacular example: it shows that sugar can be transformed into solid carbon and steam – this is why sugar is known as a carbohydrate.

Material: Small beaker (tall form), glass rod; table sugar, concentrated sulfuric acid (careful!), water, aluminum foil.

Procedure: Spread out aluminum foil on a table and place a beaker on top. Fill the beaker to half of its volume with sugar, and damp the sugar with a little water. Add sulfuric acid (volume of half of the volume of sugar) and stir intensely for a short moment. Wait.

Observation: The sugar reacts to a black matter, this matter begins to simmer, steam escapes. The porous black substance rises a few centimeters above the beaker in a toothpaste shape, could break off and falls onto the aluminum foil.

Tip: Caution: the black carbon is still moistened with concentrated sulfuric acid; it should be removed with the glass rod from the beaker, wrapped in the aluminum foil and disposed of in the solid waste container. The beaker should then be thoroughly rinsed.

E7.2 Reaction of Metals with Sulfuric Acid

Problem: Diluted acids can dissolve base metals and produce hydrogen and a salt solution. In the case of the reaction of magnesium with dilute sulfuric acid, hydrogen and magnesium sulfate solution are produced. Concentrated acids dissolve a metal too; however no hydrogen is formed. In the reaction of concentrated sulfuric acid with zinc, the gas hydrogen sulfide is produced which has a "rotten-egg" smell. It is therefore important, in the case of sulfuric acid to differentiate between the pure acid from the diluted acidic solution and should be labeled carefully. Both reactions are not acid–base reactions – they are redox reactions (see Chap. 8).

Material: Test tubes, burner; magnesium ribbon, zinc powder, concentrated sulfuric acid, about 2 M sulfuric acid solution.

Procedure: (a) Fill a test tube to one-third with sulfuric acid solution, add a 5-cm length of magnesium ribbon, and place an empty test tube of the same size on its top. Following the reaction, bring a flame close to the test tube on top to test the gas. Evaporate the water carefully from a small part of the solution in the first test tube. **(b)** Fill a second test tube to one-tenth with concentrated sulfuric acid; add a spatula tip of zinc powder. Carefully examine the odor of the produced gas.

Observation: (a) A lively bubbling development of gas begins, the solution in the test tube gets hot, the piece of magnesium decreases in size and is completely dissolved. The gas in the second test tube reacts near the flame with a special "pop" sound: hydrogen. Crystals of white salt crystallize from the solution after the evaporation of the water: magnesium sulfate. **(b)** A small amount of gas can be observed, the gas has a peculiar smell: hydrogen sulfide.

Tip: Because both reactions are classic redox reactions, they should not be interpreted within the chapter "acids and bases" – the phenomena should merely be demonstrated. The theme "redox reactions" picks up on both reactions again (see Chap. 8).

E7.3 Limestone Deposit Removers – Acidic Household Cleaners

Problem: The "removal" of limestone with an acid solution may be considered by the students as a form of "destruction" – the object of the lesson is to uncover this misconception and to point out that the "removal of calcium deposit" is a chemical reaction which forms new substances. This chemical reaction is in fact an acid–base reaction, which can be described with a proton transfer from hydronium ions to carbonate ions to form a salt solution and gaseous carbon dioxide.

Material: Beakers, test tubes; calcium carbonate powder, dilute hydrochloric acid, household limestone deposit remover, candle.

Procedure: Place a little amount of calcium carbonate powder in the beaker and cover with a few ml of hydrochloric acid. After a short while, insert a burning candle into the beaker. After reaction is over, evaporate the remaining solution. Repeat the reaction with limestone deposit removers (acidic household cleaners); check the type of acid from its label and interpret the information.

Observation: A lively gas development begins, the lit candle extinguishes in that gas: carbon dioxide. A white substance remains after evaporation: calcium chloride.

E7.4 Drain Cleaner – an Alkaline Household Chemical

Problem: Young people are familiar with acids as substances which can decompose other materials. They are usually not aware, though, that hydroxide solutions are also aggressive substances which should not be placed near the skin but especially not near the eyes. The decomposing properties are mainly used to dissolve and remove "organic leftovers" from the kitchen or bathroom. The

7.4 Experiments on Acids and Bases

aluminum chips, which can be found together with a mixture of sodium hydroxide crystals and sodium nitrate in the "drain cleaner" substance, have the purpose of forming hydrogen and ammonia which are more effective in dissolving the organic leftovers.

Material: Beakers, thermometer, glass rod; sodium hydroxide, drain cleaner, wool, paper.

Procedure: Cover several pieces of sodium hydroxide with a few ml of water in a beaker and stir with the thermometer. Add wool and paper pieces to the solution and stir well. Repeat the test with drain cleaner, and interpret the label on the bottle.

Observation: Sodium hydroxide dissolves and the mixture gets very hot, wool and paper pieces dissolve in the concentrated solution. The same happens in the drain cleaner solution.

Tip: The remains of the solution should be thoroughly rinsed with a lot of water. Even diluted amounts of hydroxide solutions are dangerous because when the water evaporates, high concentrations of sodium hydroxide are produced and cause holes in clothing.

E7.5 Reaction of Several Acid–Base Indicators

Problem: Acids as well as base solutions are mostly colorless and cannot easily be identified. Color indicators are suitable for deciding whether a solution is acidic or alkaline. Special universal indicators (in solution or paper form) can do even more: it shows the approximate pH value of a solution through various red and blue shades. The pH value can thus be introduced in a practical manner without the necessity of wondering about concentration of hydronium ions or even logarithmic calculations.

Material: Small and large beakers, test tubes, pipettes; common laboratory acid and alkaline solutions, sodium chloride solution, concentrated sulfuric acid, universal indicator solution and indicator paper (with a color chart), phenolphthalein solution.

Procedure: (a) Dilute a little amount of universal indicator solution with tap water, place the green solution in three small beakers (can be projected with an overhead projector). Drop hydrochloric acid into one of the beakers, in the other sodium hydroxide solution, in the third common salt solution. Place several drops of the solutions on strips of universal indicator paper, observe and compare the colors with the color chart, determine the pH values. **(b)** Fill a large beaker with tap water and add a few ml of phenolphthalein solution while stirring. Prepare a second beaker of the same size with a few drops of sodium hydroxide solution, a third beaker of the same size with a few drops of concentrated sulfuric acid. Pour the contents of the first beaker into the second one, the second one into the third beaker.

Observation: (a) Hydrochloric acid colors the green indicator red; sodium hydroxide solution makes it blue. The indicator paper test shows colors, and comparing with the colors of the pH scale on the color chart, pH values can be approximated in each case. (b) Colorless liquid in the first beaker immediately turns wine red by pouring into the second beaker; this solution becomes colorless when it is poured into the third beaker.

Tip: It is also possible to gain color extracts from red cabbage juice and cooked radishes, which demonstrate the change in colors by adding drops of acid and alkaline solutions.

E7.6 pH Values of Several Bathroom and Kitchen Chemicals

Problem: The scale of pH values of acids is unfortunately structured in such a way that along with an increase in concentration of hydronium ions, the pH value decreases – that for 1 M hydrochloric acid or nitric acid the pH value 0 results and all diluted solutions show pH values higher than 0. This unusual scale can be studied using universal indicator paper and the pH scale with different colors. This paper can be dipped into solutions from the kitchen, bathroom and laboratory; pH values are numerically listed in a tabular form (see Fig. 7.9).

Material: Test tubes; kitchen and bathroom chemicals, acidic and alkaline solutions from the laboratory (see Fig. 7.9), universal indicator paper.

Procedure: Estimate the pH values of the above listed chemicals using indicator paper. Record and tabulate the estimated pH values.

E7.7 Electrical Conductivity of Solutions of Acids, Bases and Salts

Problem: Students often only see formulas of acids and bases but cannot really imagine the ions in their solutions and hence develop misconceptions of molecules in solutions (see Chap. 5). In order to point out the existence of ions in acidic and basic solutions, one should test the electrical conductivity and compare the results with measurements of salt solutions.

Material: Beakers, equipment for measuring electrical conductivity and tester; 1 M solutions of strong acids and bases, 1 M NaCl solution, mineral water, tap water, distilled water.

Procedure: Measure and compare electrical conductivity of listed solutions. Dilute the solutions, test and compare with conductivity of mineral water, tap water and distilled water.

Observation: The measured values observed for strong acids and bases are higher than those of same concentrated salt solutions. Strongly diluted solutions conduct the electrical current to a smaller extent, on the one hand like mineral water, on the other hand even weaker, like tap water. Distilled water does not conduct the electric current.

7.4 Experiments on Acids and Bases 197

E7.8 Hydrochloric Acid – from Solid Sodium Chloride

Problem: Historically, hydrogen chloride gas has been produced by the reaction of sodium chloride and concentrated sulfuric acid. This reaction can be observed by students and can be interpreted as a proton transfer from H_2SO_4 molecules to Cl^- ions of sodium chloride (see Fig. 7.12). With the recognition of the colorless gas and the production of hydrochloric acid, both substances should be clearly differentiated and be appropriately described in a correct manner: the HCl(aq) solution contains the involved ions, and of course the gas HCl(g) contains molecules.

Material: Gas developer apparatus, upright cylinder with glass cover, test tubes; sodium chloride, concentrated sulfuric acid, universal indicator paper.

Procedure: Fill the gas developer under the fume hood with sulfuric acid and sodium chloride. Dropping the acid onto the solid salt develops a gas and fills the cylinder through air replacement. First check the gas with a damp universal indicator paper, place a little water and shake, test the solution with indicator paper.

Observation: The substances strongly react in the gas developer, the apparatus warms up, and a colorless gas appears. The gas forms a whitish vapor, it also has a very shocking odor. The damp indicator paper is colored bright red by the formed gas, same color appears in the solution.

Tip: If one fills a dry round flask completely with hydrogen chloride and allows water to rise into it with the aid of a glass tube from a glass bowl, then more water will be pulled up so high that it flows like a fountain into the flask: "fountain test". One liter of water reacts with up to 400 liters of hydrogen chloride to form hydrochloric acid solution. If one saturates the solution with hydrogen chloride, one gets saturated hydrochloric acid with a mass content of 37%: this is laboratory-grade concentrated hydrochloric acid.

E7.9 Hydrochloric Acid – from Gaseous Hydrogen Chloride

Problem: Hydrogen chloride reacts strongly with water and forms hydrochloric acid solution (see E7.8 "Tip"): HCl molecules react by the transfer of protons to H_2O molecules and create hydronium ions H_3O^+(aq) and chloride ions Cl^-(aq) (see Fig. 7.12). In order to follow this process in an experiment, the colors of universal indicator and the electrical conductivity should be tested and interpreted in this reaction.

Material: Glass bowl, syringe with glass tube, measuring equipment for electrical conductivity and tester; hydrogen chloride (see E7.8), water, universal indicator solution.

Procedure: Fill the syringe with hydrogen chloride (see E7.8). Fill a glass bowl up to one third of its volume with tap water, and add indicator solution until it turns green (can be projected with overhead projector). Dip the conductivity tester into the indicator solution. Add hydrogen chloride on top of the solution using the syringe (shouldn't be dipped in too deep – otherwise the fountain effect can set in, see E7.8 "Tip"). Observe electrical conductivities and color changes.

Observation: At first, the green indicator solution shows no electrical conductivity. As soon as hydrogen chloride is released, the conductivity of the solution increases and the indicator color changes from green to red.

Tip: The thermometer is dipped into water, the tip is moistened with hydrogen chloride gas (flask of the gas developer E7.8): one observes that the temperature rises up to 70°C and indicates a strong exothermic reaction.

E7.10 Sulfuric Acid and Water – a Strong Exothermic Reaction

Problem: Students know from their experiences and through dilution of liquids that the active agent is only been diluted: a very sweet sugar solution becomes diluted to a weak tasting liquid; acidic solutions like vinegar become less acidic; the same applies to many other solutions. If, however, one dilutes pure sulfuric acid with water, an intense exothermic reaction sets in: H_2SO_4 molecules transfer protons to H_2O molecules to form hydrated ions, an acid–base reaction takes place releasing a large amount of hydration energy.

Material: Beaker (small), thermometer; pure sulfuric acid, water.

Procedure: Fill a small beaker up to one-third with water. Now, carefully and slowly, add the same amount of pure sulfuric acid. Determine the maximum temperature. (Caution: add acid to water, and not the reverse, see also E7.10 "Tip").

Observation: The solution temperature rises up to 90°C.

Tip: This reaction becomes dangerous, when the pure sulfuric acid with the high density of 1.8 g/ml is used and water is added: because of the different densities both liquids do not immediately mix, the simmering water pulls sulfuric acid drops with it and splatters all over the area! The eyes should be especially protected during experiments with pure sulfuric acid!

E7.11 Reaction of Calcium Oxide and Water

Problem: The conversion reaction of quicklime or caustic lime (CaO) to slaked or hydrated lime (Ca $(OH)_2$), is historically an important process. After limestone $CaCO_3$ is heated at a temperature of about 1000°C, calcium oxide CaO is produced; this calcium oxide reacts with water to form calcium hydroxide Ca $(OH)_2$; this can be used to regenerate limestone again by the reaction with carbon dioxide forming once again calcium carbonate $CaCO_3$ ("lime circulation").

From the chemical point of view, the reaction of calcium oxide with water is a classic acid–base reaction according to the Broensted theory: oxide ions are base particles and take protons from water molecules – in this case, water molecules are acid particles. Because students have often heard the historical argument that "calcium oxide is a base" they should take into account the more contemporary viewpoint: that oxide ions are base particles.

Material: Erlenmeyer flask, thermometer; pieces of calcium oxide (fresh), water.

Procedure: Place several pieces of calcium oxide in an Erlenmeyer flask, add a little water, and measure the maximum temperature with a thermometer.

Observation: It takes about half a minute for the reaction to begin. The mixture fizzes, the beaker mists up, and the temperature rises to 100°C. The formed calcium hydroxide shows a larger volume than the original amount of calcium oxide – however both substances are white.

Tip: The calcium hydroxide can be suspended in the Erlenmeyer flask with a little water; gaseous carbon dioxide is added, and it is then closed off with a stopper and glass tube. If the flask is connected to a syringe which is filled with carbon dioxide, it is then possible to see the reaction of calcium hydroxide with the gas: the piston is sucked into the syringe. This experiment can simulate the setting and reaction of calcium mortar in the air.

E7.12 pH Values – by Dilution of Hydrochloric Acid

Problem: Students are formally aware, that a 10^{-1} molar HCl shows a pH value of 1, the 10^{-2} molar HCl has a pH value of 2. Despite this knowledge, they often cannot transfer the fact that a dilution of the factor 1:10 is necessary in order to arrive from the first to the second solution. In this experiment, they should take into account, that the change of pH values from 1 to 2 is equivalent to the dilution of 1:10, that the dilution of 1:100 results to the change of two pH units: the pH values do not relate to a linear scale, but rather to a logarithmic scale (like the Richter Scale in the measurement of earthquakes).

Material: 100 ml graduated cylinders; 0.1 M hydrochloric acid, water, universal indicator paper.

Procedure: Estimate the pH value of 0.1 M hydrochloric acid with indicator paper. Dilute 10 ml of the solution with distilled water to 100 ml solution, measure again the pH value. Dilute the solution another time by the factor 1:10, measure this pH value too.

Observation: The following pH values are obtained: 1, 2 and 3.

E7.13 Reaction of Calcium Hydroxide Solution with Carbon Dioxide

Problem: The reaction of carbon dioxide with limewater shows the well-known "milky" precipitation: white calcium carbonate is deposited. A lesser-known fact is that through the further addition of carbon dioxide, the precipitate dissolves again. Students learn through this reaction, that in tap water or mineral water, "soluble calcium" exists in the form of Ca^{2+}(aq) ions and HCO_3^-(aq) ions. The reaction shows also that calcium carbonate ("insoluble lime") can be dissolved, not only by strong acids, but also in weak carbonic acid solution.

Material: Beakers, test tubes, magnetic stirrer, measuring equipment for electrical conductivity and tester, pH meter; calcium hydroxide solution (limewater), carbon dioxide gas.

Procedure: Fill a beaker up to one-third with limewater and place it on the magnetic stirrer, which is already turned on. Dip the glass electrode of the pH meter and measure the pH value. Similarly dip conductivity tester and measure the electrical conductivity. Introduce carbon dioxide gas until the initial white precipitate fully disappears.

Observation: At first, the pH of the limewater shows a value of 12 and the electrical conductivity is high. By bubbling carbon dioxide through the limewater, the pH value goes down to 7 and low electrical conductivity is observed. When the white precipitate is deposited, nearly no conductivity is detected. Finally, adding more gas causes the precipitate to completely disappear; the pH value is about 6 and the electrical conductivity increases again.

E7.14 Neutralization Reactions

Problem: In interpreting the reaction of hydrochloric acid and sodium hydroxide solution, students have a tendency to formulate the gross equation, "HCl + NaOH → NaCl + H_2O," and even to speak of the "production of salt". When they are asked about the related ions and explain the curve of the electrical conductivity (see left diagram in Fig. 7.15), the following misconception is often heard: the electrical conductivity decreases, because four ions react in the elementary reaction, afterwards two ions are left over. In order to create appropriate concepts, the interpretation of the conductivity curve should be geared towards the fact that, during the neutralization reaction, H_3O^+(aq) ions of the acid solution are replaced by Na^+(aq) ions of the sodium hydroxide solution, that the total number of ions remains the same (see Fig. 7.14). The decrease in conductivity is discussed as being due to the varying mobility of the ions: the H_3O^+(aq) ions deliver a much larger mobility than the Na^+(aq) ions in solution.

Material: Beakers, test tubes, burette, magnetic stirrer, measuring equipment for electrical conductivity, graduated cylinder, burner; diluted hydrochloric acid solution, 0.1 M sodium hydroxide solution, phenolphthalein, universal indicator paper.

Procedure: (a) In the preliminary test, add a few drops of phenolphthalein solution to 10 ml of sodium hydroxide solution until it becomes wine-red, then add hydrochloric acid until the indicator color changes. Heat part of the solution and evaporate water. **(b)** Now, measure and compare the electrical conductivity. Pour 50 ml of hydrochloric acid in a beaker, add indicator solution and magnet, and place the beaker on the magnetic stirrer. Then place the conductivity tester in the solution, and measure the conductivity. Fill the burette with the hydroxide solution and add 1 ml of it to the acid, measure the conductivity again. Add 1 ml-portions of hydroxide solution constantly, and measure the conductivity strength until the color of the solution changes. Add several ml of hydroxide solution, determine and compare the conductivity strength, tabulate the measurements and draw a diagram.

Observation: (a) The wine-red color of the solution turns colorless, by heating this clear solution water evaporates and white crystals are formed: sodium chloride **(b)** Electrical conductivity of the hydrochloric acid solution show at first relatively high values. During neutralization it decreases to a lower level: the resulting sodium chloride solution gives the minimum of conductivity strength and the equivalent point. With an excess of hydroxide solution the conductivity values increase again (see left diagram in Fig. 7.15).

Tip: The concentration of hydronium ions can be calculated from the amount of the known concentration of the sodium hydroxide solution. The concentration of hydroxide ions can be confirmed by taking the pH value by a pH meter.

If a computer is available with all needed equipment and the conductivity tester can be connected, the conductivity curve can be obtained directly on the screen. The same is possible for pH measurements with the pH meter to obtain the curve of the pH values during titration.

E7.15 Conductivity Titration of Baryta Water with Sulfuric Acid Solution

Problem: Saturated barium hydroxide solution ("baryta water") reacts with sulfuric acid in two ways: on the one hand, hydronium ions react with hydroxide ions to form water molecules. On the other hand, Ba^{2+}(aq) ions combine with SO_4^{2-}(aq) ions to an ion lattice: solid barium sulfate is deposited. If one follows this precipitation reaction with a conductivity tester, in this particular case the current strength reverts to a value of almost zero: neither the formed water nor the deposited solid barium sulfate conducts the electricity.

Material: Beakers, burette, magnet stirrer, measuring equipment for electrical conductivity and tester; 0.1 M sulfuric acid solution, saturated barium hydroxide solution ("baryta water"), universal indicator solution.

Procedure: (a) In the preliminary test, mix small amounts of 0.1 M sulfuric acid solution and saturated barium hydroxide solution. Observe and explain the precipitation of white crystal deposit. **(b)** Place 50 ml of barium hydroxide solution in a beaker, add the indicator solution, switch on the magnetic stirrer, and dip the conductivity tester into the solution. Fill the burette with sulfuric acid, drop 1 ml of it into the hydroxide solution, and measure the conductivity. Repeat the measurements with 1-ml portions until the acid reaction of surplus sulfuric acid is clearly shown through the red color of the indicator.

Observation: The electric current decreases to zero, but increases rapidly after the color of the indicator changes from blue to red.

Tip: The concentration of barium ions and hydroxide ions can be calculated from the amount of sulfuric acid used. The concentration of hydroxide ions can be confirmed by taking the pH value of the saturated hydroxide solution. A computer may be helpful (see E7.14 "Tip").

E7.16 Heat of Neutralization

Problem: Unfortunately, the reactions of acidic and basic solutions are not spectacular: colorless solutions result again in colorless salt solutions. In order to show, at least, through temperature measurements, that an exothermic reaction of hydronium ions and hydroxide ions exists, same amounts of 2 M solutions of hydrochloric acid and sodium hydroxide solution are used: a temperature increase of approximately 13°C can be shown; a neutralization enthalpy of 57 kJ/mol will be calculated. If one chooses arbitrary pairs of strong acids and bases of the same concentration, then it can be seen that the temperature is always the same: it does not depend on the "spectator ions", but only on concentrations of H_3O^+(aq) and OH^-(aq) ions.

Material: Beakers, 100 ml graduated cylinder, thermometer; 2 M solutions of hydrochloric acid, nitric acid, sodium hydroxide and potassium hydroxide.

Procedure: Mix 50 ml of a 2 M hydrochloric acid with exactly 50 ml of 2 M base solutions. Similarly mix 50 ml of a 2 M nitric acid solution with exactly 50 ml of 2 M base solutions. Measure and compare the increase in temperatures of all the four possible mixtures.

Observation: The average increase of temperatures is measured to about 13°C.

Tip: It is also possible to mix powder of citric acid and sodium hydroxide. After heating a little bit the reaction starts: drops of water are formed, hot steam heats the test tube up to 100°C. Using these two solids it is obvious that water is produced by the neutralization.

E7.17 pH Values of Strong and Weak Acids

Problem: Students associate the term "weak acid" mostly with an acid of weak concentration, respectively with high pH values of 3, 4 or 5. They have to experience themselves through experiments, that the criterion for weak acids is not the pH value but the degree of protolysis. In order to make this criterion clear, pH value of 0.1 M hydrochloric acid could be measured and compared with the pH value of 0.1 M acetic acid solution: instead of the expected pH value of 1.0, one gets for acetic acid a pH value of 2.9. This cognitive conflict leaves an open question.

Because a pH value of about 3 exists for the 0.1 molar solution, the concentration of hydronium ions should be 10^{-3} mol/l. So this concentration is 1/100 less than the concentration of HAc molecules: only 1 of 100 HAc molecules protolyses into hydronium ion and acetate ion, 99 out of 100 molecules remain as non-protolysed molecules, the degree of protolysis is about 1%. This mental model should be transferred in the form of model drawings (see Fig. 7.16) and should be discussed in detail.

Material: Beakers, pH meter; 0.1 M solution of hydrochloric acid, 0.1 M nitric acid and 0.1 M acetic acid, universal indicator paper.

Procedure: Measure the pH of the 0.1 M solution of hydrochloric acid, 0.1 M nitric acid and 0.1 M acetic acid using indicator paper as a preliminary test. Afterwards, calibrate the pH meter for the acid range and measure the pH values of all three solutions.

Observation: Hydrochloric acid and nitric acid solution color the indicator paper red, color comparisons show a pH value of 1. Acetic acid solution colors the indicator paper light red, the pH value of about 3 is indicated. More accurate measurements with the pH meter for hydrochloric acid, nitric acid, and acetic acid show the pH values 1.0, 1.0 and 2.9.

E7.18 pH Values of HCl and H_2S Solutions

Problem: The problem of weak acids can also be estimated by using gases. If one fills a syringe with 24 ml hydrogen chloride and a second one with 24 ml hydrogen sulfide gas, one knows that each has 1 mmol molecules at room temperature: both gas samples contain the same amount of about 6×10^{20} molecules. If they are both dissolved in the same amount of water, approximately both in 10 ml of water, one obtains same concentrations of 0.1 molar solutions. Because of both 10^{-1} molar acidic solutions, one could expect a pH value of 1 in both cases. While a pH value of 1 can be measured by the HCl solution, a pH value of 4 is observed with the H_2S solution. The cognitive conflict can be solved by taking into account that hydrogen sulfide shows only partial protolysis with the protolysis level of 0.1%.

Material: Beakers, two syringes, pH meter; hydrogen chloride gas, hydrogen sulfide gas, water, universal indicator paper.

Procedure: Fill one syringe with 24 ml of hydrogen chloride gas, and the other one with 24 ml of hydrogen sulfide gas. Add 10 ml of water in both syringes and shake until the gas samples are dissolved. Measure the pH values of the solutions using an indicator paper and a pH meter.

Observation: The indicator paper is colored red with the hydrochloric acid solution, but not with the second solution. The pH meter shows a pH value of 1 and 4, respectively.

Tip: The hydrogen sulfide solution can be used to precipitate certain metal solutions and to show that insoluble deposits are produced: metal sulfide precipitation for cation analysis!

References

1. Strube, W.: Der historische Weg der Chemie. Von der Urzeit zur industriellen Revolution. Leipzig 1976 (Deutscher Verlag für Grundstoffindustrie)
2. Asselborn, W., u.a.: Chemie heute, Sekundarbereich II. Hannover 1998 (Schroedel)
3. Christen, H.R., Baars, G.: Chemie. Frankfurt 1997 (Diesterweg Sauerländer)
4. Musli, S.: Säure Base Reaktionen: Empirische Erhebung zu Schuelervorstellungen und Vorschlaege zu deren Korrektur. Münster 2004 (Examensarbeit)
5. Barker, V.: Beyond Appearances – Students' Misconceptions About Basic Chemical Ideas. London 2000 (Royal Society of Chemistry)
6. Sumfleth, E.: Schülervorstellungen im Chemieunterricht. MNU 45 (1992), 410
7. Geisler, A., Sumfleth, E.: Veraenderung von Schuelervorstellungen im Bereich Saeuren und Basen. In: Brechel, R.: Zur Didaktik der Physik und Chemie. Alsbach 1999 (Leuchtturm)
8. Barke, H.D.: pH neutral oder elektrisch neutral? Über Schuelervorstellungen zur Struktur von Salzen. MNU 43 (1990), 415
9. Sumfleth, E.: Ueber den Zusammenhang zwischen Schulleistung und Gedaechtnisstruktur. Eine Untersuchung zu Saeure-Base-Theorien. NiU PC 21 (1987), 29
10. Schmidt, H.J.: A label as a hidden persuader: Chemists' neutralization concept. International Journal of Science Education 13 (1991), 459
11. Schmidt, H.J.: Harte Nuesse im Chemieunterricht. Frankfurt 1992 (Diesterweg)
12. Schmidt, H.J.: Students' Misconceptions – Looking for a Pattern. SciEd 81 (1997), 123
13. Fruehauf, D., Tegen, H.: Blickpunkt Chemie. Hannover 1993 (Schroedel)
14. Asselborn, W., u.a.: Chemie heute, Sekundarbereich I. Hannover 2001 (Schroedel)
15. Dehnert, K., u.a.: Allgemeine Chemie. Hannover 1993 (Schroedel)

Further Reading

Cros, D., Maurin, M., Amouroux, R., Chastrette, M., Leber, J., Fayol, M.: Conceptions of first-year university students of the constituents of matter and the notions of acids and bases. European Journal of Science Education 8 (1986), 305

Hand, B.M.: Students understanding of acids and bases: A two year study. Research in Science Education 19 (1989), 133

Hand, B.M., Treagust, D.F.: Application of a conceptual conflict strategy to enhance student learning of acids and bases. Research in Science Education 18 (1988), 53

Hawkes, S.J.: Arrhenius confuses students. Journal of Chemical Education 69 (1992), 542

Ross, B., Munby, H.: Concept mapping and misconceptions: A study of high school students' understandings of acids and bases. International Journal of Science Education 13 (1991), 11

Schmidt, H.-J.: A label as a hidden persuader: Chemists' neutralization concept. International Journal of Science Education 13 (1991), 459

Fig. 8.1 Concept cartoon concerning oxidation and reduction

Chapter 8
Redox Reactions

Today, redox reactions are defined as electron transfer reactions. For lectures in lower grades, we also use the historic understanding concerning the transfer of "oxygen" or "O atoms", later in lectures of advanced courses, this historic definition will be expanded to include the electron transfer representation.

At first, the historic reduction term has nothing in common with today's concept of the redox reaction. It can be traced back to the German scientist Joachim Jungius or Junge (1587–1657) who described the metamorphosis of ore to the pure metal as a *reduction* [1]. Through intense heating, it was usual in Jungius' time, to reduce cinnabar (solid red quicksilver sulfide) to mercury; thereby large amounts of liquid mercury have been extracted and filled, for example, in the decorative wells of Spanish rulers.

The oxidation term was later developed after the discovery of oxygen. C.W. Scheele, who had already discovered oxygen in 1771, described this gas as "fire air". Priestley, a strong advocate of the phlogiston theory, gave oxygen the name "dephlogisted air" [1]. Probably, Lavoisier owes his appreciation to Scheele and Priestley for their non-critical adherence to the phlogistone theory and for the fact, that he took scales for measuring masses of oxygen. Therefore, he is considered the real discoverer of oxygen.

Lavoisier placed a large emphasis in his chemical laboratory on a correct methodology, on exact measurements and note taking. At first – looking for an experimental confirmation of Stahl's phlogiston theory – he was able to relay, that the increase in weight during the burning of metals was exactly the same amount of weight which was lost in the sealed equipment. Via the reverse reaction – a thermolysis of mercury oxide – he was finally able to show that this consists of two elements: mercury and oxygen. With this, he took over Boyle's element definition; he had, however, a continuing agenda. He differentiated between the *matèrie* and *principe: matèrie,* according to Lavoisier, is visible; *matèrie sulfurique* – sulfuric material – is yellow and melts to a light liquid. Sulfuric acid, in comparison, combines the *principe sulfurique* and *oxygènique:* the principle which creates acids. The new element is known as *oxygènium* (Lat.-Greek: acid-forming), because he assumed that oxygen is to be found in all acids [1]. After all, it was Lavoisier who coined the term oxidation, to indicate reactions in which elements react with oxygen to form oxides.

Initially, the terms oxidation and reduction were not at all opposites: reduction stood for the extraction of a metal from metal compounds, and oxidation for the reaction with oxygen. However, with time, the meaning of reduction changed more in favor of a reaction, in which oxygen is released. With this, reduction and oxidation have become terminologically paired – to two types of chemical reactions, of which one represents the reverse of the other.

With Thomson's discovery of electrons in 1897 and the development of concepts concerning electron loss and gain of atoms or ions, the term **oxidation and reduction** has changed to the **loss of electrons** and **gain of electrons** changed at the beginning of the 20th century.

However, with this definition came a considerable change in meaning. While the oxygen definition is based on the oxidation of matter, for example that of metal-to-metal oxides, the definition of electron transfer is based on the smallest particles: the Fe atom is oxidized under the release of two electrons to an Fe^{2+} ion. In a similar way, other atoms or ions are reduced through the addition of electrons. Therefore, a correspondence of oxidation and reduction always exists and one speaks of reduction–oxidation reactions or **redox reactions**:

$$\text{Fe atom} \rightarrow Fe^{2+} \text{ion} + 2\,e^- \qquad \textbf{oxidation}$$

$$2\,\text{Cl atoms} + 2e^- \rightarrow 2\,Cl^- \qquad \textbf{ions reduction}$$

$$Fe + Cl_2 \rightarrow Fe^{2+}(Cl^-)_2 \qquad \textbf{redox reaction}$$

Again, extended definitions could be based on **oxidation numbers** of atoms in molecules. An oxidation is, in this regard, a process in which the electronic surroundings of an atom in a molecule or in an ionic compound changes – i.e. in such a way, that polar bonds are formed, whereby especially O atoms extract neighboring electrons in their surrounding. By definition, oxidation numbers change as in the following example:

$$\overset{0}{S_8} + 8\,\overset{0}{O_2} \rightarrow 8\,\overset{+IV\;-II}{SO_2}$$

The oxidation number of an S atom in the S_8 molecule is defined as 0, it increases to $+IV$ for an S atom in the SO_2 molecule. Formally, the reduction of O atoms in O_2 molecules takes place to O atoms in SO_2 molecules, the oxidation number decreases from 0 to the value $-II$. The roman numerals, which are used for oxidation numbers, should be reserved for these formal charges of atoms in molecules or ions; arabic numerals are designed as charge numbers of ions.

The same applies as before: the oxidation numbers do not relate to the substances, but to particles. In this regard, certain atoms in a molecule show

Table 8.1 Redox definitions [2]

Definition	Oxidation	Reduction
1	Gain of oxygen	Loss of oxygen
2	Loss of electrons	Gain of electrons
3	Increase of oxidation number	Decrease of oxidation number

specific oxidation numbers: the C atom shows an oxidation number of $-$IV in the CH_4 molecule, the C atom in the CO_2 molecule shows the oxidation level of $+$IV. The three definitions of oxidation and reduction are summarized in Table 8.1 [2].

Also, in the case of acid–base reactions, the acids are, at first, defined at a substance level. However, according to Broensted's theory, the smallest particles can give protons or can take protons, the particles are acids or bases. The same difficulties arise: students cannot always successfully decide if the macroscopic level of the substances or the sub-microscopic level of the particles is intended. Misconceptions in this and other matters should be described and clarified.

8.1 Misconceptions

As in the historical development of the redox concept, the definition of oxygen transfer in beginners' lessons changes to electron transfer in advanced lessons as soon as the subject of atomic structure is discussed. After the introduction of the oxygen definition, there is a belief that oxygen has to be involved in every redox reaction. The reason for this is, first of all, the syllable –ox, which is semantically strongly associated with the name oxygen (*oxygenium*, oxide), and with combustion reactions in air or oxygen which are carried out in the introductory lessons.

Schmidt [3] described appropriate studies. In these studies, almost 5000 students had to decide on which of his listed reactions belonged to redox reactions: the reaction of dilute hydrochloric acid with (1) magnesium, (2) magnesium oxide, and (3) magnesium hydroxide. We know of course that (1) is to be identified as a redox reaction, (2) and (3) as acid–base reactions: in (2) H_3O^+(aq) ions react with O^{2-} ions of the oxide, and in (3) H_3O^+(aq) ions react with OH^- ions of the hydroxide.

Approximately half of the students in advanced courses and one-third of the basic courses chose the correct answer. The remaining students marked the oxygen-related answer and gave explanations like: "(2) and (3) contain oxygen, which is absolutely necessary for redox reactions (12th grade, advanced course); oxygen is necessary for every redox reaction, so (1) cannot be a redox reaction (11th grade, basic course); (2) and (3) are redox reactions because in both cases oxygen and electron transfer takes place (11th grade, basic course)" [3]. Elsewhere Schmidt cites this student's comment: "Oxidation means: a reaction in which oxygen is involved. The ending 'oxide' shows that (2) as well as (3) are redox reactions" [4].

The oxygen concept seems so strong that Schmidt [2] tells this story about a typical acid–base reaction: "Garnett and Treagust, in 1992, asked senior high school students whether or not the equation

$$CO_3^{2-} + 2H^+ \rightarrow H_2O + CO_2$$

represents a redox reaction. All students with the correct answer used the oxidation number method. Those who answered incorrectly had two reasons. One was to assume that the carbonate ion had donated oxygen to form carbon dioxide and was, therefore, reduced. The other was to assign the oxidation number to polyatomic species by using their charge. CO_3^{2-} was given the oxidation number of negative 2, and CO_2 the oxidation number of zero. Consequently, the reaction '$CO_3^{2-} \rightarrow CO_2$' was identified as an oxidation. In a similar manner, the reaction "$H_3O^+ \rightarrow H_2O$" can be identified as a reduction: the hydronium ions must have gained electrons in the transformation from hydronium to water molecules, and so should have been reduced" [2].

Sumfleth [5] asked students grades 6–12 in Germany to provide an explanation regarding the popular reaction of an iron nail in copper sulfate solution. She documented incorrect answers, which could be traced back to preconcepts, as well as school-made misconceptions. Especially, students in grades 6–8 (15% of all students) described the formation of a copper-colored coating with "sedimentation, clinging to, sticking to, or color fading of a material on an iron nail" or "the copper sulfate colors the iron nail, the copper sulfate sticks on to it, like when a piece of wood is placed in a dye and is then dried". Of course – these explanations are based on everyday observations [5]. Half of the 7th grade students guessed "an attraction of the substances" as the reason, the other students mentioned a pre-existing magnetism – probably because of the iron nail. These students however, only described their observations with words, one cannot admonish them for their preliminary ideas.

Even in senior high school classes, these discussions remain – they are merely peppered with specialized terminology: "copper peels away from sulfate and is mounted on the iron; copper sulfate is reduced; copper atoms attract electrons; iron nails can absorb ions from the solution" [4]. These statements demonstrate that individual subject terminology can be learned and that students feel the urge to use them. However, it appears as if they are coincidentally interspersed in order to give the appearance of some sort of scientific background.

Further tests by Sumfleth [5] and Stachelscheid [6] were association tests; due to many answers, the same misconceptions were found. The interesting thing was the frequency with which individual terms and correlating explanations were given. A third of those questioned explained the term redox reaction purely tautologically: "a redox reaction is an addition of reduction and oxidation; consists of an oxidation and a reduction" [6]. In the association test, the term oxidation was used much more often (58%) than reduction (43%) or redox reaction (41%) and was associated with oxygen. In association with combustion,

8.1 Misconceptions

students also mention oxidation (21%) much more often than reduction (3%) or redox reaction (5%). Again, the terms oxidizing agents and reducing agents are mainly tautologically explained (15% of students). Only 6% of the students report that oxidizing agents gain electrons and that they themselves are reduced and vice versa [6].

Many other references show misconceptions in the area of redox reactions, especially with the interpretation of voltage and electric current in electrolysis or Galvanic cells. Therefore, Marohn [7] looked for the mental models that students develop by discussing Galvanic cells. In addition, Garnett and Treagust discovered conceptual difficulties in the area of electric circuits [8] and electrolytic cells [9], the same with Ogade and Bradley working on electrode processes [10], Sanger and Greenbowe investigating common students' misconceptions in electrochemistry [11] or current flow in electrolyte solutions and the salt bridge [12]. Nevertheless, these topics are so difficult to understand that misconceptions can hardly be avoided – especially concerning the nature of electrons as waves and/or particles, concerning the electromagnetic fields and their forces. Therefore, the emphasis here is to look to the basic definitions of the redox reaction and to discuss common experiments that students can perform themselves.

Empirical Research. In the following paragraphs, our own results of empirical research are presented. With the help of a questionnaire, Vitali Heints [13] carried out new studies in German high schools where redox reactions have been introduced as electron transfer. Several problems are based on preconcepts regarding combustion which have already been mentioned in Chap. 3. However, in Problem 1 of the questionnaire, information about the combustion of copper is asked while, in Problem 2, it is compared to the reaction of copper with chlorine.

Reactions of Metals with Oxygen and Chlorine. With this problem, we tried to get information about how students describe their concepts of metal–oxygen reactions: as oxygen transfer or as electron transfer, and how the chosen definition is applied to the metal–chlorine reactions. The other point of interest is to determine if the chosen definition is discussed by oxidation or reduction of *substances* or of *particles*.

Problem 1 A sheet of thin copper is folded to a little envelope and heated with the hot flame. The red copper turns black on the outside, on the inside the red color remains. This is because:

- ❏ a combustion reaction is taking place   [A]
- ❏ black soot is deposited on the outside   [B]
- ❏ a redox reaction is taking place   [C]
- ❏ copper atoms change their color   [D]

The second distracter [B] was the most attractive choice and was chosen by 59% of all students. The correct answer [C] was chosen 21%, the distracters [A] and [D] were chosen by 18% and 4% of students, respectively. The popularity of distracter [B] can be traced back to a lack of practical chemical work experience.

It seems that most students do not know differences between a "yellow flame" and a "blue hot flame". The majority of the explanations for [B] read roughly as follows: "soot is formed or created through the burning flame or the burning fire". In addition, the students are not clear about the role of oxygen in combustion processes, they look for everyday explanations: "oxygen is burned and soot is deposited; oxygen is burned and carbon dioxide is created; the combustion of oxygen can lead to the products CO, CO_2, C".

Problem 2 A thin sheet of copper is heated and placed in a gas jar which is filled with yellow chlorine gas. The copper glows and a green substance appears because:

- ❏ copper reacts with chlorine [A]
- ❏ chlorine forms hydrochloric acid and the metal is destroyed [B]
- ❏ an acid–base reaction takes place [C]
- ❏ chlorine destroys copper atoms [D]

The answer which was chosen most was in fact the correct one [A], that is 61%. However, 7% of students chose another distracter, especially [D] in addition to [A], hence we can say that only 54% answered it. The distracters [B], [C] and [D] were chosen 10%, 4% and 3%, respectively. The explanation given by one student in an advanced class is exemplary and reads: "There is no O_2, so that no CuO can be formed, nevertheless a redox reaction takes place, but only with chlorine". Despite the knowledge of the extended definition, an electron transfer was not added to the interpretation, even the advanced students prefer to work with the more familiar oxygen definition – the extended definition is only used when there is no obvious connection to oxygen.

Corrosion. Due to the fact that corrosion plays a large role in industry and society, this topic is a meaningful subject in chemistry lessons and problems. It is especially important to analyze the way students use simple and extended redox.

Problem 3 Iron does not rust in dry California; however, in Germany, it rusts very easily because the air contains quite a lot of moisture. This is because:

- ❏ iron contains rust which shows up when exposed to air [A]
- ❏ iron atoms become oxidized [B]
- ❏ an acid–base reaction takes place [C]
- ❏ iron atoms are destroyed through rust [D]

Students' preconcepts are particularly addressed through the well-known formation of rust, conflicts between everyday language and modern redox theory readily occur. Luckily, 56% of the students chose the correct answer [B]; the distractors [D], [C] and [A] were chosen by 11%, 10% and 3%, respectively. Twenty-five percent of those questioned gave no answer. Many explanations however, indicate a basic initial knowledge, because the expected electron transfer did not play any role but rather, one chose to work much more with simple oxygen interpretations: "rust is oxidized iron; iron is oxidized; iron or iron atoms combine with oxygen". The reaction of iron with water or steam is

8.1 Misconceptions

rarely mentioned: "iron attracts humidity from the air and oxidizes with it; iron atoms oxidize with the oxygen from the steam; iron has no protective layer and is attacked by the humidity".

Furnace Process. Students should know the process of producing the most important metal: iron. They like to say "by heating the ore, iron melts and seeps out". So we will inform them that ore has to be mixed and heated with coal and hope that a redox reaction can be easily determined.

Problem 4 Iron is produced by mixing iron ore (i.e. iron oxide, Fe_3O_4) and carbon in a furnace and heating both substances until glowing iron flows out. In this process:

- ❏ carbon is a catalyst [A]
- ❏ a redox reaction takes place [B]
- ❏ iron oxide is reduced [C]
- ❏ iron oxide is decomposed into the elements [D]

40% of the students did not even try to work on this problem. The answer most often marked was "carbon is a catalyst" (26%); the answers "reduction of iron oxide" and "decomposition into the elements" were each recorded with about 21%. The correct answer [B] was marked by 20% of the students. Only a few students of the secondary level gave correct explanations complete with reaction equations. No one described the redox reaction of iron oxides with carbon monoxide formed by burning the coal.

Many of their explanations are quite elementary: "carbon reacts with oxygen to CO_2, Fe remains; oxygen is needed to burn carbon and this is taken from iron oxide; the solid material iron gets its stability through carbon; carbon only helps in getting the reaction started, does not react; however; carbon supplies the heat, which is necessary for splitting the oxide".

It appears that students avoid the electron transfer definition because the oxygen definition is more familiar. This definition has been learned in chemistry lessons using words in simple reaction equations. It has also made an impact due to the related experiments with sound and fire effects that can easily be recalled. The extended redox definition has perhaps been formally worked out on the blackboard but not necessarily intensified so that the electron transfer didn't cause a "conceptual change" in the cognitive structure.

Reaction of Metals with Solutions. The following problems can no longer be interpreted with the oxygen definition, but necessitate the concept of electron transfer. These reactions are easy to demonstrate through experiments and can especially be used as students' experiments – this is why one assumes that each student knows these phenomena and is able to interpret them.

Problem 5 An iron nail is dipped into a copper sulfate solution. After some time, a copper-colored coating can be found on the nail. Explain the observation.

Altogether, about 34% of the students used the term redox reaction or oxidation–reduction and electron transfer (22% of those from the beginner level, 66% from the advanced level); 14% use the terms "precious metal or base metal" to characterize the phenomena. Reaction equations are often used but they are not always correct. The majority of answers are formed using the same accurate pattern: "copper is more precious than iron; because copper forms a coating on the nail, Cu^{2+} ions are reduced from the copper sulfate so that Cu is formed, a redox reaction took place".

In many cases, students do not differentiate between copper and copper sulfate and copper sulfate solution, or between substances and particles: "copper ions from the solution connect to the iron nail". Several students assume the metal coating is rust: "the copper-colored matter is rust; iron is attacked which results in rust; the nail rusts after being dipped into the copper sulfate solution". Others do not see any reaction in the process and assume attracting forces or magnetic interactions: "the nail is magnetic and attracts the sulfate; copper from the solution is magnetic, just like the nail".

22% of the students assume a "rubbing-off or sedimentation or clinging of elemental copper from the copper sulfate solution, of copper atoms, of copper ions, of copper electrons". Some write "copper sulfate solution is saturated and combines with iron atoms of the nail; the copper electrons cling to the nail and a coating is formed; the copper sulfate is reduced to copper and iron; copper ions from the solution combine with the iron nail; the iron nail reacts with the copper sulfate solution, the iron absorbs electrons from the $CuSO_4$ and forms a copper color through a redox reaction; copper coats the iron nail, hence the copper-colored coating, the active iron enters the solution and the copper ions form a precipitation".

The evaluation of the data shows that approximately half of the ordinary level students attributed the precipitation of copper by using familiar everyday explanations. These explanations are at times so dominant, that even the explicitly named copper color is not taken into account: "iron comes into contact with the copper and colors it green; the copper-colored coating is green-turquoise-blue". In these cases, copper is associated with the color green of well-known "copper roofs" which are really coated by a special copper carbonate compound.

School-made misconceptions play a more significant role by advanced students, they do not differentiate between atoms and ions, atoms and substances, or substances and particles; they are simply used as synonyms. Several students do not seem to be aware that redox reactions between atoms and ions are necessary to explain the formation of elemental copper – in their mind copper sulfate solution already contains the copper atoms: "copper from copper sulfate forms a deposit on the iron nail, it connects to the iron atoms". Students definitely lack an understanding of the concept of ions in salt solutions (see Chap. 5).

Problem 6 Magnesium (Mg) reacts with hydrochloric acid, HCl (aq), under the development of a gas, the gas is verified as hydrogen (H_2):

❏ a redox reaction takes place	[A]
❏ an acid–base reaction takes place	[B]
❏ chlorine particles are oxidized	[C]
❏ magnesium atoms are oxidized	[D]

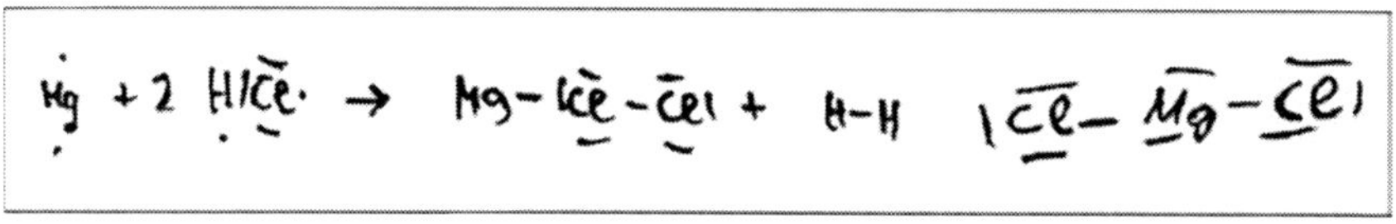

8.1 Misconceptions

> $\overset{..}{Mg} + 2\ H/\overline{Ce}\cdot\ \rightarrow\ Mg-\overline{Ce}-\overline{Ce)} +\ H-H\quad |\overline{Ce}-\overline{Mg}-\overline{Ce)}$

Fig. 8.2 Molecules as mental models for particles in acid and salt solutions [13]

Both correct answers [A] and [D] are the most attractive with 38%, and 19%, respectively; both distracters [B] and [C] are only marked by 10% and 8%, respectively. The problem is considered solved when either [A] or [D] is chosen and sufficiently explained. Explanations should contain at least one appropriate reaction equation. As an example, several appropriate explanations are: "magnesium is active, active metals react with acids and hydrogen is formed; Mg is oxidized; Mg and HCl combine to form $MgCl_2$". Examples of incorrect explanations are: "magnesium is oxidized and hydrochloric acid is reduced; both materials are altered, magnesium absorbs electrons and hydrochloric acid emit electrons; the H atom splits from Cl and combines to H_2; Mg and Cl react to $MgCl_2$". In many cases, inappropriate molecule symbols were given instead of using ionic symbols of the involved ions (see Fig. 8.2). Also answers like "the electron transfer is responsible for this result" are provided without any further comments which make the response incomplete and meaningless.

Problem 7 A copper strip and a magnesium strip are inserted in a lemon in such a way, that they do not touch each other. A voltmeter is connected to both metal pieces; a voltage of approximately 1 V is measured:

- ❏ the acid in the lemon is responsible for the voltage   [A]
- ❏ an acid–base reaction takes place   [B]
- ❏ a redox reaction takes place   [C]
- ❏ metal atoms are going in solution   [D]

Only 60% of students worked on this problem; 40% of the answers were attributed to distractor [A], 21% and 9% to distractors [B] and [D], respectively. Only 19% of the students chose the correct answer [C]. Because the correct choice was often combined with the marking of a different distractor, the real amount of correct answers was only 7%.

The majority of answers and explanations for distracter [A] come from students of the ordinary level and are limited to statements such as "acids conduct electricity" and to everyday analogies: "acid from a lemon is the same as battery acid; the lemon conducts the voltage because the lemon's acid is electrically charged; copper and magnesium, along with citric acid, release energy". Students who made these and similar statements mainly chose [A] alone or [A] combined with the distracter [B] which suggests that the acidic lemon reaction is an acid–base reaction.

Those who tended to concentrate on both metals in their explanations mostly chose a combination of answers [C] and [D]: "this is an electrolysis; the atoms are capable of moving from one strip to the other because of the acid; the solution pressure of Mg is higher than the precipitation pressure of Cu; Cu is more noble than Mg, therefore electron transfer".

The reason for the low number of correct answers mostly lies in the misconception that the mere presence of acids automatically leads to an acid–base reaction. Even when a student has understood the redox reaction and has supplied a plausible explanation, misconceptions still remain, as is shown in the following example: "Cu is nobler and is therefore reduced, it accepts electrons and Mg is oxidized because the active metal releases electrons, the voltage is created through ion formation".

Oxygen and Oxidation Numbers. Redox definitions and oxidation numbers must finally come into play here in order to find students' mental models appropriately.

Problem 8 When a substance is oxidized:

❏ it takes up oxygen   [A]
❏ it takes up electrons   [B]
❏ it works as an oxidizing agent   [C]
❏ it works as a reducing agent   [D]

The correct answers [A] or [D] were given by 49% and 24%, respectively; 9% of students chose answer [B] and likewise answer [C]. A total of 40% of the answers were correctly given – all of these students chose answer [A] and [D]. The explanations of the beginner level students were limited to statements such as "oxidation = oxygen absorption" or "oxidation = due to oxygen". Explanations of advanced level students follow the pattern: "oxidation = electron release = reducing agent".

One answer is particularly interesting: "in the oxidation process, a substance loses electrons which are then replaced by oxygen". This pupil appears to be attempting to combine the definition based on electron transfer with the oxygen definition in order to explain his decision for choosing [A] and [D]. Furthermore, it is not clear what the oxidized material withdraws from the electrons – the "substitution of electron through oxygen" does not make one conclude that the electrons are absorbed by oxygen and by O atoms, respectively. Such a statement, once again, indicates the existence of interim misconceptions.

So we conclude that students from the beginner level are already using the expanded definition of redox reactions, at least as much of the oxygen definition in the inductory lessons. However, both definitions are constantly being mixed and it appears as if they have never been taught the particle-related electron transfers. This also applies to the majority of the advanced students, who use an expanded definition when answering these questions but rarely state from where the electrons come and where the electrons are being transferred. It is apparent that these students have already been introduced to the terms electron transfer and electron release.

8.2 Teaching and Learning Suggestions

The following Problems 9 and 10 are only given to students of the advanced level because they are the only ones who have already dealt with oxidation numbers in class.

Problem 9 Which of the following conversions describes an oxidation?

- ☐ Mn^{2+} → $MnCl_2$ [A]
- ☐ MnO_4^- → MnO_4^{2-} [B]
- ☐ $Mn(OH)_2$ → MnO_2 [C]
- ☐ Mn^{3+} → Mn_2O_3 [D]

Problem 10 Which equation correlates to a reduction?

- ☐ $Cr^{3+} + 3e^-$ → $Cr(s)$ [A]
- ☐ Cr^{3+} → $Cr(s) + 3e^-$ [B]
- ☐ $Cr(s)$ → $Cr^{3+} + 3e^-$ [C]
- ☐ $Cr(s) + 3e^-$ → Cr^{3+} [D]

The correct answer to Problem 9 is option [C], it was chosen by 45% of the students. Only 28% of the answers were selected as correct because students supplied acceptable explanations: "oxidation numbers are increased; because the oxidation number is more positive; the oxidation number of Mn increased from +II to +IV; electrons are gained".

In Problem 10, the most commonly chosen answer was [A] which is correct, but only half of the students marked this answer. A few of them supply an explanation: "Cr accepts e^-; electrons are taken up; reduction = uptake of electrons". Explanations are mostly limited to the mere naming of terms and a restating of the oxidation numbers and no comments. It is especially apparent that most students are not clear about "+ $3e^-$" as to whether it should appear on the left or on the right side of the reaction equation. Students do not notice the charge differences between the right and the left side of the equation.

Summary. It is determined that the use of the oxygen definition is very common among students; they use it whenever possible. Most students tend to use everyday language in order to describe redox reactions. Chemical terms at the substance level are arbitrarily mixed with those at the level of the smallest particles; molecules are used instead of ions for salt or acid solutions. Redox equations and electron transfers are rarely correctly described or explained.

8.2 Teaching and Learning Suggestions

In Sect. 3.5, misconceptions about the topic of **combustion** have already been analyzed and experiments for a better understanding of reactions involving air and oxygen have been suggested. These recommendations can also be discussed with respect to the definition of redox reactions as an oxygen-transfer (see E3.1, E3.2, E3.8 and E3.10). However, initially, writing those reactions with word equations like copper oxide(s) + iron → copper(s) + iron oxide; exothermic

and explaining these as reduction of copper oxide and oxidation of iron would be better than pointing out the famous topic of "oxygen-transfer" from copper oxide to iron. The latter would definitely contribute to the development of well-known misconceptions regarding the "release of free oxygen". Even if an equation is written using formulas, then the transfer of "O atoms" – or of "O_2 molecules" would result. The students should know that the oxide ions are retained and neither an O atom nor an O_2 molecule is transferred; the redox reaction takes place exclusively between Cu^{2+} ions and Fe atoms through electron transfer. For this reason, it appears to be more advantageous not to use equations with formulas for redox reactions, especially when mentioning the "oxygen" definition, but rather to formulate these reactions in words.

Mentally, a lot is expected from students when advancing from the oxygen definition to the electron transfer one, particularly as the oxidation and reduction has to be interpreted at the level of the smallest particles. In this sense, the reduction is no longer limited to a metal oxide as a substance, but rather to the reduction of a metal ion to the metal atom. Because this mental effort, the "conceptual change" is not achieved, the related misconceptions should, if possible, be prevented or corrected.

Electron Transfer. The most dramatic mistakes made by students are caused by the interference of the oxygen concept with the electron concept. Although the electron definition has been dealt with for quite some time, one 12th grade student writes, after discussing the reaction of hydrochloric acid with magnesium oxide and magnesium hydroxide, that "both substances contain oxygen, which is absolutely necessary for a redox reaction" [3].

This is why these interferences should probably be excluded by not comparing the metal–oxygen reactions to the metal–chlorine reactions, but by immediately starting with the metal precipitations from solution: they can exclusively be interpreted by electron transfer.

A prerequisite for the interpretation of **metal precipitations** is the term "ion" and the simple structure of the atom with the atom nucleus and differentiated electron shells. If the ion term has already been introduced as in Chap. 5, then the colors of salt solutions are already known – for instance the light blue color of diluted copper sulfate solutions or of diluted copper chloride solutions. Armed with this information, there are good prerequisites for the problem-oriented interpretation of the following experiments. If an iron nail is dipped into copper sulfate solution, then a copper-colored coating appears on the part that has been dipped (see E8.1). If iron wool is placed in copper sulfate solution, then the solution warms up and the blue color of the solution disappears (see E8.2).

The discoloration of the solution almost forces an interpretation, that Cu^{2+}(aq) ions from the solution "disappear", or have reacted. The question of their whereabouts leads to the supposition that they have deposited as Cu atoms on the iron and have formed copper crystals. This conjecture, if applicable, is

8.2 Teaching and Learning Suggestions

more open for discussion after the following reaction, which should preferably be carried out as a student exercise.

If a helix-shaped copper wire is placed into silver nitrate solution and one waits a few minutes, then the development of silver crystal needles can be observed and the change in the color of the initially colorless solution to blue (see E8.3). With this reaction one observes that Cu^{2+}(aq) ions appear and that copper has partially dissolved. From this reaction, one concludes that, with experiences gathered from the first experiment (see E8.1), metal atoms go into solution as ions, accompanied by the release of electrons. Along with this, metal cations of the salt solution are taking these electrons, forming metal atoms and crystallizing in a large number of atoms to a piece of metal or a needle of silver:

$$Cu \text{ atom} \rightarrow Cu^{2+}(aq) \text{ ion} + 2e^-; \quad 2 Ag^+(aq) \text{ ions} + 2e^- \rightarrow 2 Ag \text{ atoms}$$

Describing these half reactions, it should be made apparent to the students that the term "$+ 2e^-$" should be placed on the correct side of the equation: one Cu atom can only become one Cu^{2+} ion if it simultaneously releases two electrons. It is advisable to suggest to the students that the number of atoms and the number of charges should be the same "left and right of the arrow". In the given examples, the number of the charges is zero in each case.

Precipitation Sequence. It should be concluded that the ions from the more noble metals are changed into atoms and crystallized from the solution. Simultaneously, due to electron transfer, the atoms of the active metal go into solution through the formation of ions. This hypothesis can systematically be tested with other metal pairs (see E8.4); the observations are noted by the precipitation sequence of the metals.

The conversion of metal compounds to pure metals is historically known as reduction; the reduction of metal ions with the gaining of electrons is thereby explained:

$$2 Ag^+(aq) \text{ ions} + 2e^- \rightarrow 2 Ag \text{ atoms} \quad \textbf{gain of electrons, reduction}$$

The gained electrons stem from the reacting metal atoms, which simultaneously form corresponding ions by losing electrons:

$$Cu \text{ atom} \rightarrow Cu^{2+}(aq) \text{ ion} + 2e^- \quad \textbf{loss of electrons, oxidation}$$

Altogether, an electron transfer takes place from Cu atoms to Ag^+ ions:

$$Cu + 2 Ag^+(aq) \rightarrow Cu^{2+}(aq) + 2 Ag \quad \textbf{electron transfer, redox reaction}$$

The term oxidation can now be associated with well-known metal–oxygen reactions; even in these oxidations, metal atoms are transformed into their corresponding metal ions. If needed, the oxygen interpretation can be added later while discussing precipitation reactions. From this point on, all redox reactions should be consequently interpreted with electron transfers. The net result is that the named interference of the oxygen transfer and of the electron transfer definitions are minimized and that the electron transfer becomes a permanent mental model.

Redox Sequence. Even the most well-known reactions of metals with diluted acid solutions to form salt solutions and hydrogen gas can also be interpreted as redox reactions or electron transfers (see E8.5): metal atoms are oxidized, H^+(aq) ions of acid solutions are reduced to H atoms, these combine to form H_2 molecules and the hydrogen gas is released in the form of small bubbles. Because copper and the noble metals are not soluble in diluted acids, the redox pair H^+/H is placed in a location before the copper pair in the Activity Series of Metals. The name *metal series (sequence)* can be replaced by the general name *redox series (sequence)*. A partial listing is shown as follows:

Na^+(aq)/Na; Mg^{2+}(aq)/Mg; Zn^{2+}(aq)/Zn; Fe^{2+}(aq)/Fe; **H^+(aq)/H**;

Cu^{2+}(aq)/Cu; Ag^+(aq)/Ag

Voltage Sequence. In the following experiments, electron loss and gain will be observed through voltage measurements between two metals in electrolyte solutions (see Fig. 8.3). Various metal strips are dipped into rock salt solution and corresponding voltages are measured with the help of a voltmeter (see E8.6); the further apart the metals are in the redox sequence, the higher the measured voltages are in these electric cells. For instance, using a zinc plate and a copper plate, a voltage of approximately 1 V is measured. It is even more spectacular to use the acidic solution in a lemon: "electric voltage by a lemon" (see E8.6).

The measured voltages can be explained through the equilibria on both metal surfaces (see Fig. 8.3): Zn atoms of the active metal zinc, for example, release electrons on the metal surface, leave the surface of the metal, and go into solution as Zn^{2+}(aq) ions:

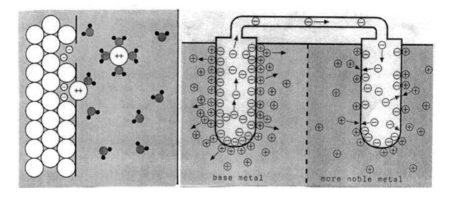

Fig. 8.3 Mental models on the formation of potential differences of two metals [14, 15]

$$\text{Zn atom} \leftrightarrows \text{Zn}^{2+}(\text{aq}) \text{ ion} + 2e^-$$

The zinc electrode is negatively charged because the equilibrium of Cu atoms and Cu^{2+} ions lies more on the side of the Cu atoms of the more noble metal copper (see Fig. 8.3). So, automatically, the copper plate becomes the positive pole in the electric cell; an electrical voltage can be measured between both metals. One can state that every metal has a specific *electrical potential*, but when present by itself, is not possible to measure. With the voltmeter, it is possible to measure the *potential difference* between two metals dipped in salt or in acidic solutions.

There is another way to explain the measured potential difference. If you postulate that every metal has different *electron pressure* and that the electron pressure of zinc is higher than copper: the voltage results from these differences of electron pressures. So, the whole *metal sequence* can be explained by a high electron pressure of active metals like magnesium and zinc, and a low pressure of noble metals. The high electron pressure of zinc or magnesium means that a piece of zinc or magnesium dissolves by donating electrons to the particles in the air (O_2 molecules, CO_2 molecules) or to particles of rain water (H_2O molecules, CO_2 molecules). The low electron pressure of gold explains the existence of that metal over thousands of years: gold does not give electrons to gases in the air, to rain water or to any other solution of the nature. These differences in the behavior of the metals are obvious when explained with electron pressure – but it is hard to illustrate these differences using models or drawings.

Electrical Current. After measuring the voltage between both electrodes of an electric cell, both metals (magnesium and copper, for example) can be connected with a small electrical motor or by a light bulb (see E8.6): the motor runs and the bulb lights up if the electric current is high enough. The excess of electrons of magnesium is transferred to the copper plate, the electric current travels from magnesium to copper. By the transfer of electrons, both equilibria of metal atoms and ions are shifted and have to rearrange: i.e. magnesium ions go into solution, producing a new excess of electrons, and the voltage stays the same: the electric current can flow through. It should be made clear that, initially, an electric voltage has to build up before an electrical current will flow.

In precipitation reactions or redox reactions in solution, the electrons always go directly and not measurably from one atom or ion to another one. The new thing about redox reactions in electric cells is that a measurable electrical charge is built up and after a consumer is switched on, the electron flow runs through a metal cable: the electrons are not directly exchanged but initially form an electrical current. This transmits electrical energy which can be made visible to the students in the form of a running motor or a lighted bulb. In this connection, the initial historical experiences with electric voltage and electric current by Galvani and Volta in the 18th century can be presented, compared and discussed.

Galvanic Cells. In the initial experiments with electric cells, the measured values depend on the concentration and electrode distances and are therefore not reproducible. It is therefore advisable to measure the electrical voltage between two standardized half-cells (see Fig. 8.4). In using solutions of salts which are 1 molar in concentration of the metal ions, one speaks of standard conditions: in the related example (see Fig. 8.4) the standard voltage of 1.1 V (see E8.7) can be measured.

The voltage can be measured by the differences of electron pressure or by different equilibria between metal atoms and ions. The Zn/Zn^{2+} equilibrium is more shifted towards the ions, the Cu/Cu^{2+} equilibrium is more shifted towards the atoms, forming the different poles:

Negative pole: $Zn(S) \leftrightarrows Zn^{2+}(aq) + 2e^-$ **electron excess on metal**

Positive pole: $Cu^{2+}(aq) + 2e^- \leftrightarrows Cu(s)$ **electron deficiency on metal**

Zn^{2+} ions go from the zinc electrode into solution, each ion leaving behind two electrons and forming an excess of electrons, the negative pole of the Galvanic cell. Cu^{2+} ions touch the copper electrode, react to Cu atoms by the electrons of the copper metal and form copper crystals. Due to the resulting electron deficiency, a positive pole is hereby formed.

For the formation of equilibria it is necessary that the anions in the Galvanic cell – in this case the sulfate ions – can diffuse through the permeable diaphragm, for example through the clay cell from unglazed clay (see E8.7): they move into the zinc half-cell and balance out the electrical charge of the newly formed zinc ions (see Fig. 8.4). The sulfate ions are diffused from the copper half-cell because they no longer need to compensate for the charge of those Cu^{2+} ions which have reacted to form Cu atoms and copper crystals on the metal.

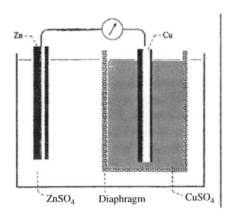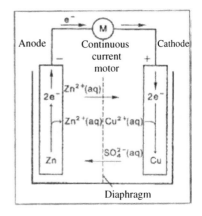

Fig. 8.4 Mental model of the Galvanic cell $Zn/Zn^{2+}(aq) // Cu^{2+}(aq)/Cu$ [16]

Further standard voltages could be measured and placed into an electrochemical sequence with other pairs of half-cells, as can be found in many chemistry teaching books. The extent of the introduction of the hydrogen half-cell as a standard electrode can be determined in each individual lesson. If one consequently describes all redox reactions in relation to the metal sequence, redox or electrochemical sequence with ions and offers the students model drawings (see Figs. 8.3 and 8.4), then the electron transfer and the redox definition, in terms of involved smallest particles, becomes even clearer and the mixing at the language level of substances and that of particles can be effectively suppressed.

By formulating redox equations, it is especially necessary to discuss half reactions for the oxidation and reduction process. If one takes a look at the number of exchangeable electrons and thereby at the total sum of the charges "left and right of the arrow", the students do not only have a control instrument at hand for testing the correctness of the equations, but also gain a better understanding for such redox reactions.

Oxidation Numbers. There are many redox reactions which should not be interpreted in the sense of electron transfer, but rather with a **shifting of electrons.** The most obvious example is the exothermic hydrogen-oxygen reaction to form water:

$$\overset{0}{2\,H_2} + \overset{0}{O_2} \rightarrow \overset{+I\ -II}{2\,H_2O}$$

In this example, such a shift of electrons could be expressed in such a way that from non-polar electron-pair bonds in H_2 molecules and O_2 molecules one gets polar bonds in H_2O molecules. Also, from non-polar molecules dipole-molecules are formed. This process can be described in a model with imaginary charge numbers, the oxidation numbers. They supply the number of charges which are contained in an atom of a compound when one imagines that the compound is composed of ions. For this formal deliberation, the connective electron pairs are completely arranged to the bonding partner with the higher electronegativity.

In the mentioned example, the oxidation number 0 of the H atoms in H_2 molecules is increased to the value of $+I$ in H_2O molecules. In addition, one correspondingly finds the reduction of O atoms in O_2 molecules to O atoms in H_2O molecules; the oxidation number is lowered from 0 to $-II$. In this sense, every oxidation takes place with an increase of the oxidation number; every reduction with the decrease in the oxidation number.

It is possible to describe reactions using the model of oxidation numbers, as in the example of the reaction of zinc with pure sulfuric acid. By using diluted acids, one gets hydrogen; with pure sulfuric acid, zinc metal forms the nasty-smelling hydrogen sulfide (see E7.2):

Oxidation half reaction: $Zn \leftrightarrows Zn^{2+}(aq) + 2e^-$

Reduction half reaction: "$H_2SO_4 \leftrightarrows H_2S$"

The reduction of S atoms, with the oxidation number $+VI$ in H_2SO_4 molecules to S atoms with the oxidation number $-II$ in H_2S molecules takes place with the gaining of 8 e^- for which 4 Zn atoms are necessary per elementary step. In addition, four O atoms of the H_2SO_4 molecule should be transformed with 8 H^+ ions into 4 H_2O molecules:

Oxidation: $4\ Zn \leftrightarrows 4\ Zn^{2+}(aq) + 8\ e^-$

Reduction: $H_2SO_4 + 8\ e^- + 8\ H^+ \leftrightarrows H_2S + 4\ H_2O$

Redox reaction: $4\ Zn + H_2SO_4 + 8H^+ \leftrightarrows 4\ Zn^{2+}(aq) + H_2S + 4\ H_2O$

With this description, it becomes apparent why exactly 4 Zn atoms per elementary step have to react: it is necessary to have a balance of lost and gained electrons of the oxidation and of the reduction step. In this case, in the first step of the reaction, 8 electrons are necessary, 4 Zn atoms have to lose them. Apart from this, one should check, if the charges "left and right of the arrow" are equal in the oxidation and in the reduction step, and may equate the complete equation. Since the hydrogen sulfide gas escapes from the equilibrium, a shift in favor of the products occurs.

Corrosion. With the help of oxidation numbers, there are further important redox reactions that can be described, for instance the chemical processes in the corrosion of iron. If one shows in an experiment that iron only goes rusty when water is present in addition to air (see E8.8), then it is possible to recognize the corrosion as a redox reaction:

Oxidation: $2\ Fe(s) \leftrightarrows 2\ Fe^{2+}(aq) + 4\ e^-$

Reduction: $2\ H_2O + O_2 + 4\ e^- \leftrightarrows 4\ OH^-(aq)$

Redox reaction: $2\ Fe + 2\ H_2O + O_2 \leftrightarrows (Fe^{2+})_2(OH^-)_4$ (s, reddish brown)

By the precipitation of iron hydroxide the equilibrium shifts in favor of further rust formation. But rust is not pure iron hydroxide – it is a mixture of oxides and hydroxides of iron.

Batteries. Processes in batteries and car lead accumulators can now be examined and clearly interpreted. In a Leclanché battery, there is a zinc container, which represents the negative pole. The positive pole is formed by a graphite rod, which is surrounded from a mixture of carbon and manganese dioxide (see Fig. 8.5).

8.2 Teaching and Learning Suggestions

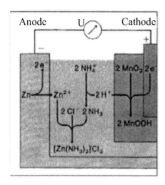 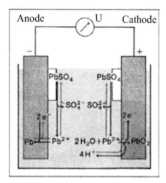

Fig. 8.5 Mental models on processes in Leclanché batteries and lead accumulators [16]

In a model battery, a zinc plate is dipped into a 20% ammonium chloride solution, a graphite rod with manganese dioxide represents the other pole, a voltage of approximately 1.5 V can be detected (see E8.9). The processes in the battery can be understood as follows:

$$\text{Negative pole: } Zn(s) \leftrightarrows Zn^{2+}(aq) + 2e^-$$

$$\text{Positive pole: } 2\,MnO_2(s) + 2\,H^+(aq) + 2\,e^- \leftrightarrows 2\,MnOOH(s)$$

$$\text{Redox reaction: } Zn(s) + 2\,MnO_2(s) + 2\,H^+(aq) \leftrightarrows Zn^{2+}(aq) + 2\,MnOOH(s)$$

In the reduction process, the oxidation number of the two Mn atoms decreases from +IV to +III, so that they need just two electrons for the first step. One Zn atom delivers the two electrons from the oxidation step, the equilibrium leads to a voltage of approximately 1.5 V. As soon as an electrical consumer is connected, the equilibrium shifts in favor of the particles to the right side of the equations. A model drawing demonstrates the chemical reactions (see Fig. 8.5).

Even the reactions in the lead accumulator can be interpreted with the help of oxidation numbers (see Fig. 8.5). If, in the model of the accumulator, two lead plates are dipped into 20% sulfuric acid solution and electrolyzed by the voltage of about 5 V, one of the lead plates is covered with a dark layer of lead oxide (see E8.10). If the transformer is taken away and one joins both plates with a voltmeter, the voltage of 2 V will be measured (see E8.10). The following equilibrium exists if a pure lead plate and a lead oxide plate are separately dipped into 20% sulfuric acid solution (see Fig. 8.5):

$$\text{Negative pole: } Pb(s) \leftrightarrows Pb^{2+}(aq) + 2e^-$$

$$\text{Positive pole: } PbO_2(s) + 4\,H^+(aq) + 2\,e^- \leftrightarrows Pb^{2+}(aq) + 2\,H_2O$$

$$\text{Redox reaction: } Pb(s) + PbO_2(s) + 4\,H^+(aq) \leftrightarrows 2\,Pb^{2+}(aq) + 2\,H_2O$$

If electrons are withdrawn from the negative pole, the equilibrium supplies them once again and the reactions run from left to right when discharging the accumulator. If one wishes to recharge the batteries by setting up a special voltage, then the reactions run from right to left. The processes of discharging and charging are as follows (see Fig. 8.5):

$$\text{Pb(s)} + \text{PbO}_2(\text{s}) + 2\,\text{H}_2\text{SO}_4(\text{aq}) \underset{\text{discharging}}{\overset{\text{charging}}{\rightleftarrows}} 2\,\text{PbSO}_4(\text{s}) + 2\,\text{H}_2\text{O}$$

Further examples for redox reactions could be possibly shown through more special cell batteries and fuel cells, industrial electrolyses could also be interesting in this case.

Summary. It is pointed out that, in order to avoid misconceptions, the introduction of ions is very important: ions have been dealt with as basic particles of matter according to Dalton's atomic model (see Chap. 5). In order to understand the charges of ions and the change of ions and atoms by electron transfer, the differentiated atomic model with nucleus and electron shells should be introduced. With the assistance of a clear terminology, it is easy to formulate half-reaction for the oxidation and reduction steps, the number of electrons to be transferred can be clearly recognized. Finally, if mental models – for instance, from involved atoms or ions in Galvanic cells or in batteries – are relayed and drawn by the students themselves, then they could more easily see through the redox processes or even perhaps be able to repeat them independently. In all explanations, one should pay attention that the observations should be done at the substance level, but that the interpretations and discussions of reaction equations should consequently take place at the level of the smallest particles as atoms, ions and molecules.

8.3 Experiments on Redox Reactions

E8.1 Precipitation of Copper from Copper Sulfate Solution

Problem: Student's misconceptions often show that the oxygen definition for the redox process is mixed with that of the electron transfer: these mixed concepts are neither apt, nor are they educationally helpful. Because concepts on the historical "oxygen-transfer" are often reproduced from introductory lessons, it is necessary, in advanced lessons, to immediately begin with appropriate explanations based on electron transfer. One example is the precipitation of metals from their solutions ("cementation").

Material: Test tubes; iron nail, copper sulfate solution.

8.3 Experiments on Redox Reactions

Procedure: Fill a test tube one-third with copper sulfate solution; dip an iron nail half into the solution. Wait for a minute, take out the nail from the solution and scratch the red deposit on the nail with a spatula.

Observation: The part of the iron nail submerged in the solution has a red coating: copper. This coating can be removed; unchanged iron is easily visible after the coating has been removed.

E8.2 Exothermic Precipitation Reactions

Problem: If, in addition to the iron nail experiment, the reaction of iron wool or magnesium powder with the blue copper sulfate solution is shown, the heat of reaction of the exothermic reaction can be determined. On the other hand, the blue solution discolors in this reaction so that, with the "disappearance" of Cu^{2+}(aq) ions, the discussion of the formation of Cu atoms from the appropriate ions could be expanded.

Material: Test tubes, thermometer; iron wool, magnesium powder, copper sulfate solution.

Procedure: Fill two test tubes to one-third with copper sulfate solution. Add iron wool to the first test tube and magnesium powder to the second one. Determine the temperature before and after both reactions by hand or using a thermometer.

Observation: The solutions get quite warm; the metals are covered with a red coating: copper. The color of the solution changes from blue to colorless: mixture of copper sulfate and silver sulfate solution.

E8.3 Precipitation of Silver from Silver Nitrate Solution

Problem: It has been recognized in the first example, that the precipitation of relatively noble copper is no coincidence. With this additional example of silver precipitation, it should be confirmed, that from two participating metals, the more noble metals precipitate and the active metals go into solution. For this reason, copper wire is dipped into silver salt solution, and the silver is placed into copper salt solution: both reactions should be compared.

Material: Test tubes, glass rod; copper wire, copper sulfate solution, silver wire, silver nitrate solution.

Procedure: Coil a piece of copper wire in helix form using glass rod, take the glass rod away, hang the helix-shaped copper wire in silver nitrate solution and observe. Repeat this test with silver wire and copper sulfate solution.

Observation: Silver needles are formed on the copper wire; the colorless salt solution gains a light blue color: copper nitrate solution. No reaction takes place in the second experiment.

E8.4 Precipitations with Other Metals

Problem: With the help of the previous experiments, further suppositions occur as to which metals react with which metal salt solutions. They can be tested in test tubes; afterwards the metals are arranged in the well-known metal activity sequence.
   Material: Test tubes; various metal strips or wires, appropriate salt solutions.
   Procedure: Dip the metal pieces systematically into different salt solutions and test for reactions. Record positive reactions in tabular form.
   Observation: Relative noble metals like copper, nickel and lead are deposited from their solutions in the presence of active metals like magnesium, zinc or iron.

E8.5 Metals in Acidic Solutions

Problem: Diluted acids dissolve active metals and are accompanied by hydrogen gas formation: metal sulfate solutions are formed in reactions with diluted sulfuric acid, metal chloride solutions in the case of reactions with hydrochloric acid. Because acids are involved, one tends to jump to the conclusion that it has to be an acid–base reaction (see E7.2). In order to make clear that one is dealing with redox reactions, the following experiments should be interpreted in that sense.
   Material: Test tubes; magnesium ribbon, zinc granules, diluted sulfuric acid and hydrochloric acid solution.
   Procedure: Fill a test tube to one-third with acid solution, add a 3-cm magnesium ribbon or a zinc granule. An empty upside-down test tube of the same size is placed on top of the first test tube. Following the reaction, a flame is placed near the mouth of the test tube on top. Finally, carefully evaporate water from the first test tube.
   Observation: A lively gas development starts, the solution in the test tube becomes hot, and the piece of metal becomes smaller and completely dissolves in the acid. The gas collected in the test tube gives a pop sound when tested with a flame: hydrogen. After water is evaporated from the solution a white salt crystallizes: magnesium sulfate or zinc chloride.

E8.6 Metals in Salt Solutions

Problem: In the previous reactions, electrons are directly transferred from one particle to the other – an electric current cannot take place. If the electrons are diverted back from one metal electrode to the other electrode, both dipped into a salt solution, it is possible to measure, first the voltage with a voltmeter or second to demonstrate the current using a light bulb or an electric motor. Recorded voltages are not normalized because the distances between electrodes or solution concentrations are arbitrary. However, it is possible to guesstimate their relative position in the electrochemical series based on the values obtained.

Material: Glass Beaker, voltmeter (multimeter), cable and alligator clips, light bulb, electric motor; various metal strips or rods, sodium chloride solution.

Procedure: Fill a beaker two-thirds with salt solution, dip a copper strip to one side of the beaker and magnesium ribbon to the other side. Attach both metals to a multimeter, record the voltages. Check formation of an electric current with a light bulb or an electric motor. Replace magnesium by other metals like zinc or iron, or even with copper; record the values measured.

Observation: For the metals copper and magnesium, a voltage of about 1.7 V is recorded. The electric motor starts to move and runs for a while. The other voltages are smaller; the motor no longer runs. If the same metals are combined, then no voltage is measurable.

Tip: In order to attain a spectacular "voltage out of a lemon", two different metal strips could be placed inside the lemon and the voltages measured: the juice of the acidic cell serves as an electrolyte solution in this case. If the voltage or the current is too weak to start the motor or the light bulb, the multimeter can be used to measure the electric current.

E8.7 Galvanic Cells

Problem: The measurements in the previous experiments are not standardized, voltage values of the same metal pairs can vary according to the conditions. If half-cells are produced by dipping the metal pieces in their corresponding 1 M salt solution, and in each case two half-cells are connected together, the result is a Galvanic cell. Using the metal pair copper–zinc, a standard voltage of 1.1 V is always obtained. The larger the metal pieces and surfaces are chosen, the greater the electric current strength is attained for a Galvanic cell.

Material: Beaker, unglazed clay cell, voltmeter (multimeter), electric motor, cable and alligator clips; copper strip and 1 M copper sulfate solution, zinc strip and 1 M zinc sulfate solution.

Procedure: Fill a beaker to half with copper sulfate solution. Fill a clay cell with zinc sulfate solution. Dip metal strips into their respective solutions. The clay cell which contains zinc sulfate solution is placed into the beaker that contains copper sulfate solution. Connect both metal strips to the multimeter; finally attach an electric motor into the electric circuit.

Observation: A voltage of 1.1 V is measured, the electric motor runs.

Tip: In case no clay cell is available, it is sufficient to prepare the half-cells in two beakers and to connect them to a strip of filter paper which has been dipped into sodium nitrate solution (salt bridge). The half-cells can also be exchanged: a magnesium or nickel half-cell can be connected to the mentioned copper half-cell; voltages can be determined.

E8.8 Iron Corrosion

Problem: Students are aware of metal–oxygen reactions and are of the impression that rusting of iron is a pure iron–oxygen reaction. However, if one recalls from experience that, in the Californian or Sahara deserts oxygen exists but that cars never rust, then it can be assumed that the rusting of cars in Europe has to be connected to the damp air prevalent in the climate. The following experiment verifies the hypothesis that oxygen as well as water are necessary for iron corrosion.

Material: Test tubes, glass beaker; iron wool, water.

Procedure: Fill one test tube to one-third with dry iron wool, and another test tube with damp iron wool. Place both test tubes upside down in a beaker which is half filled with water. Observe the water level in both test tubes after one or two days.

Observation: No changes are observed in the test tube with dry iron wool. In the test tube with damp iron wool, the water level rises approximately 2 cm in the test tube, iron wool turns black.

E8.9 Leclanché Battery

Problem: The Galvanic cell (see E8.7) demonstrates the function of a battery. However, copper is too expensive to be considered as a battery metal. With zinc, graphite and manganese dioxide, the Frenchman Leclanché developed a battery, which is successfully used throughout the world. The zinc cup forms the negative pole and functions also as a container for the battery which is usually surrounded by a steel coat in order to avoid corrosion. A graphite electrode surrounded by manganese dioxide is placed into ammonium chloride suspension and forms the positive pole of the battery, a voltage of approximately 1.5 V is attained.

Material: Beaker, voltmeter (multimeter), electric motor, cable and alligator clips, filter paper and adhesive tape; zinc plate, graphite rod, manganese dioxide powder (MnO_2), ammonium chloride solution.

Procedure: Fill a beaker two-thirds with the salt solution, attach a zinc plate on one side with the help of a clip. Place a graphite rod at the middle of the beaker and surround it with manganese dioxide (take filter paper and adhesive tape). First connect the graphite rod and the zinc plate to a voltmeter, then with the electric motor.

Observation: A voltage of approximately 1.5 V is measured, the electric motor runs.

E8.10 Lead Accumulator

Problem: Young people are exposed to regular and rechargeable batteries almost everyday – they use them in calculators, radios, toys, etc. They may know about the lead battery in their parent's car. In order to understand how they work, an initial exposure to ordinary batteries is useful (see previous experiment). A model experiment should demonstrate the function of the lead accumulator, which is made rechargeable by connecting it to an electrical transformer. It has to be pointed out that the car battery is recharging by the generator of the car, when the motor is running and the car is moving.

Material: Beaker, battery (4.5 V), voltmeter (multimeter), electric motor, cable and alligator clips; two lead strips, sulfuric acid solution (20%).

Procedure: Fill a beaker two-thirds with sulfuric acid. Place lead strips at opposite ends and hold in place with alligator clips. Electrolyze the sulfuric acid with a battery of 4.5 V supply. Wait for about one minute, remove the battery. Measure the voltage using a voltmeter, then connect with an electric motor.

Observation: While the electrolysis is running, a gas develops on both sides of the lead plates, thereby turning one of the lead strips dark brown. After this, a voltage of approximately 2 V can be measured, the electric motor works for a short while and then stops.

Tip: Lead accumulators can be charged over and over again. If an original car battery is available, it can be dismantled and compared with the model experiment: the original one will contain six cells, which supply a total of $6 \times 2 \text{ V} = 12 \text{ V}$.

References

1. Wolter, H.: Johann Joachim Becher zum 300. Todesjahr. PdN-Chemie 31 (1982), 45
2. Schmidt, H.J.: Shift of meaning and students' alternative concepts. International Journal of Science Education 25 (2003), 1409
3. Schmidt, H.J.: Der Oxidationsbegriff in Wissenschaft und Unterricht, Chem. Sch. 41 (1994), 6
4. Schmidt, H.J.: Students' misconceptions – looking for a pattern. Science Ed 81 (1997), 123
5. Sumfleth, E., Schuelervorstellungen im Chemieunterricht, MNU 45 (1992)
6. Sumfleth, E., Stachelscheid, K., Todtenhaupt, St.: Redoxreaktionen in der Sekundarstufe I. NiU-Chemie 2 (1991), 77
7. Marohn, A.: Falschvorstellungen von Schuelern in der Elektrochemie – eine empirische Untersuchung. Dissertation. Dortmund 1999
8. Garnett, P.J., Treagust, D.F.: Conceptual difficulties experienced by senior high school students of electrochemistry: Electric circuits and oxidation-reduction equations. Journal of Research in Science Teaching 29 (1992), 121
9. Garnett, P.J., Treagust, D.F.: Conceptual difficulties experienced by senior high school students of electrochemistry: Electrochemical (Galvanic) and electrolytic cells. Journal Research of Science Teaching 29 (1992), 1079
10. Ogade, A.N., Bradley, K.H.: Electrode processes and aspects relating to cell EMF, current, and cell components in operating electrochemical cells. Journal of Chemical Education 73 (1996), 1145
11. Sanger, M.J., Greenbowe, T.J.: Common students' misconceptions in electrochemistry: Galvanic, electrolytic and concentration cells. Journal Research of Science Teaching 34 (1997), 377
12. Sanger, M.J., Greenbowe, T.J.: Common students' misconceptions in electrochemistry: Current flow in electrolyte solutions and salt bridge. Journal of Chemical Education 74 (1997), 819

13. Heints, V.: Redoxreaktionen: Empirische Erhebung zu Schuelervorstellungen und Vorschlaege zu deren Korrektur. Staatsexamensarbeit. Muenster 2005
14. Gloeckner, W., Jansen, W., Weissenhorn, R.G.: Handbuch der experimentellen Chemie. Sekundarbereich II. Band 6: Elektrochemie. Koeln 1994 (Aulis)
15. Christen, H.R., Baars, G.: Chemie. Frankfurt 1997 (Diesterweg, Sauerlaender)
16. Asselborn, W., Jaeckel, M., Risch, K.T.: Chemie heute, Sekundarbereich II. Hannover 1998 (Schroedel)

Further Reading

Ahtee, M., Varjola, I.: Students' understanding of chemical reaction. International Journal of Science Education 20 (1998), 305

Andersson, B.: Pupils' explanations of some aspects of chemical reactions. Science Education 70 (1986), 549

Barral, F.L., Fernandez, E.G.-F.: Secondary students' interpretations of the process occurring in an electrochemical cell. Journal of Chemical Education 69 (1992), 655

Bou Jaoude, S.B.: A study of the nature of students' understanding about the concept of burning. Journal of Research in Science Teaching 28 (1991), 689

Boulabiar, A., Bouraoui, K., Chastrette, M., Abderrabba, M.: A historical analysis of the Daniell Cell and elektrochemistry teaching in French and Tunisian textbooks. Journal of Chemical Education 81 (2004), 754

Clerk, D., Rutherford, M.: Language as a confounding variable in the diagnosis of misconceptions. International Journal of Science Education 22 (2000), 703

De Jong, O., Acampo, J., Verdonk, A.: Problems in teaching the topic of redox reactions: Actions and conceptions of chemistry teachers. Journal of Research in Science Teaching 32 (1995), 1097

Faulkner, L.R.: Understanding electrochemistry: some distinctive concepts. Journal of Chemical Education 60 (1983), 262

Garnett, P.J., Treagust, D.F.: Conceptual difficulties experienced by senior high school students of electrochemistry: Electric circuits and oxidation–reduction equations. Journal of Research in Science Teaching 29 (1992), 121

Garnett, P.J., Treagust, D.F.: Conceptual difficulties experienced by senior high school students of electrochemistry: Electrochemical and electrolytic cells. Journal of Research Science Teaching 29 (1992), 1079

Garnett, P.J., Treagust D.F.: Implications of research on students understanding of electrochemistry for improving science curricula and classroom practice. International Journal of Science Education 12 (1990), 147

Huddle, A.H., White, M.D., Rogers, F.: Using a teaching model to correct known misconceptions in electrochemistry. Journal of Chemical Education 77 (2000), 104

Morikawa, T., Willamsen, B.E.: Model for teaching about electrical neutrality in electrolyte solutions. Journal of Chemical Education 78 (2001), 934

Niaz, M.: Facilitating conceptual change in students' understanding of electrochemistry. International Journal of Science Education 24 (2002), 425

Ogude, N.A., Bradley, J.D.: Ionic conduction and electrical neutrality in operating electrochemical cells. Journal of Chemical Education 71 (1994), 29

Ogude, N.A., Bradley, J.D.: Electrode process and aspects relating to cell EMF, current, and cell components in operating electrochemical cells. Journal of Chemical Education 73 (1996), 1145

Öskaya, A.R.: Conceptual difficulties experienced by prospective teachers in electrochemistry: Half-cell potential, cell potential, and chemical and electrochemical equilibrium in Galvanic cells. Journal of Chemical Education 79 (2002), 735

Runo, J.R., Peters, D.G.: Climbing a potential ladder to understanding concepts in electrochemistry. Journal of Chemical Education 70 (1993), 708

Sanger, M.J., Greenbowe, T.J.: Common student misconceptions in electrochemistry: Galvanic, electrolytic, and concentration cells. Journal of Research in Science Teaching 34 (1997), 377

Sanger, M.J., Greenbowe, T.J.: Students' misconceptions in electrochemistry: Current flow in electrolyte solutions and the salt bridge. Journal of Chemical Education 74 (1997), 819

Sanger, M.J., Greenbowe, T.J.: Analysis of college chemistry textbooks and sources of misconceptions and errors in electrochemistry. Journal of Chemical Education 76 (1999), 853

Sanger, M.J., Greenbowe, T.J.: Addressing student misconceptions concerning electron flow in aqueous solutions with instruction including computer animations and conceptual change strategies. International Journal of Science Education 22 (2000), 521

West, A.C.: Electrochemical cell conventions in general chemistry. Journal of Chemical Education 63 (1986), 609

Fig. 9.1 Concept cartoon concerning aqua copper complexes

Chapter 9
Complex Reactions

The third large group of donor–acceptor reactions are the complex reactions. If we consider the bright-blue colored solution of copper sulfate in water and add concentrated hydrochloric acid, the blue color changes to green. The blue solution contains complex ions called hexaaquacopper complexes with the symbol $[Cu(H_2O)_6]^{2+}$(aq). The structure of the complex shows an octahedron (see Fig. 9.2): the Cu^{2+} ion is called central ion, 6 H_2O molecules are the ligands. Through the reaction with chloride ions the complex releases one water molecule and a chloride ion replaces it in the complex:

$$[Cu(H_2O)_6]^{2+}(aq, blue) + Cl^-(aq) \rightarrow [CuCl(H_2O)_5]^+(aq, green) + H_2O$$

If one adds water, the green solution changes back into the known blue solution, the monochloropentaaqua complex disappears, the hexaaqua complex is formed again: a chemical equilibrium exists between both complexes, different stabilities of complexes must be considered and can be calculated using the stability constants.

The stabilities of copper aqua complexes are low. Therefore, experts have two descriptions of the complex: $[Cu(H_2O)_4]^{2+}$(aq) because four water molecules form a relatively stable square around the copper ion, and $[Cu(H_2O)_6]^{2+}$(aq) because

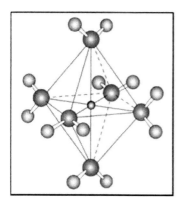 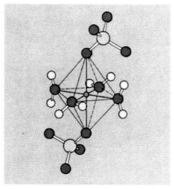

Fig. 9.2 Structure of aquacopper complexes in solution and in the solid salt [1]

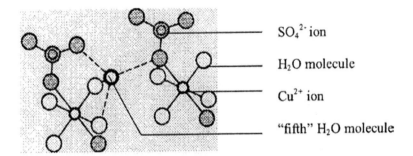

Fig. 9.3 Environment of the "fifth H_2O molecule" in $CuSO_4 \cdot 5\ H_2O$ [2]

two more water molecules are unstably bonded in the structure of a distorted octahedron (see Fig. 9.2). Both descriptions are possible: the tetraamminecopper complex is shown as $[Cu(NH_3)_4]^{2+}$(aq) or as $[Cu(NH_3)_4(H_2O)_2]^{2+}$(aq).

If one heats the blue copper sulfate solution until the dry blue copper penta hydrate salt ($CuSO_4\ 5\ H_2O$) remains, the complex is still in the ionic lattice, this time, the central ion with four H_2O ligands and two O atoms from surrounding sulfate ions (see Fig. 9.2). "The fifth H_2O is not attached to the cation, it is held between water molecules attached to cations and O atoms of sulfate ions" [2] (see Fig. 9.3). Furthermore, if one heats the blue salt the crystals turn into a white substance: H_2O molecules leave the complexes and an ionic lattice remains with the arrangement of copper ions and sulfate ions.

History. Many different structures concerning the complexes have been tested over the years. The most famous theory was the chain theory [3]: scientists postulated that the ligands are attached at the metal particle like in a chain (see Fig. 9.4). Alfred Werner was the acclaimed chemist who solved the problem in 1893 [4]: he placed the metal ion in the center of the new compound and assumed that the dichlorotetraammincobalt complex has two octahedral structures. He

	Formula	Structure
1.	$CoCl_3 \cdot 6NH_3$	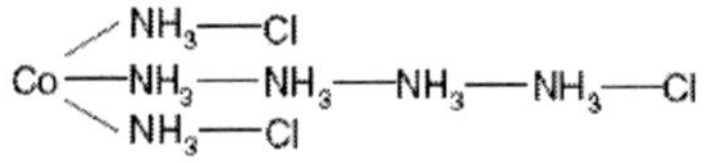 Co—NH$_3$—Cl / NH$_3$—NH$_3$—NH$_3$—NH$_3$—Cl \ NH$_3$—Cl
2.	$CoCl_3 \cdot 5NH_3$	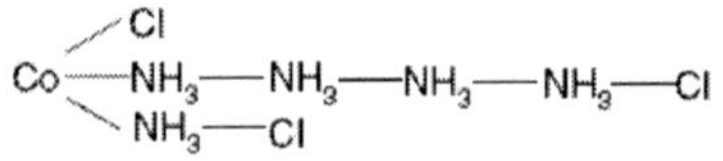 Co—Cl / NH$_3$—NH$_3$—NH$_3$—NH$_3$—Cl \ NH$_3$—Cl
3.	$CoCl_3 \cdot 4NH_3$	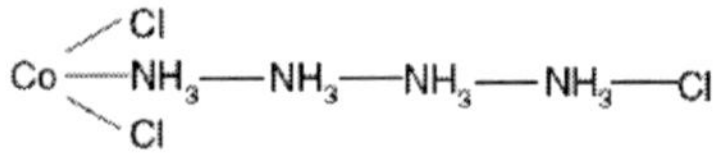 Co—Cl / NH$_3$—NH$_3$—NH$_3$—NH$_3$—Cl \ Cl
4.	$CoCl_3 \cdot 3NH_3$	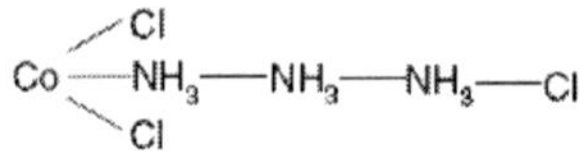 Co—Cl / NH$_3$—NH$_3$—NH$_3$—Cl \ Cl

Fig. 9.4 Chain theory of complex structures proposed by Jorgensen [3]

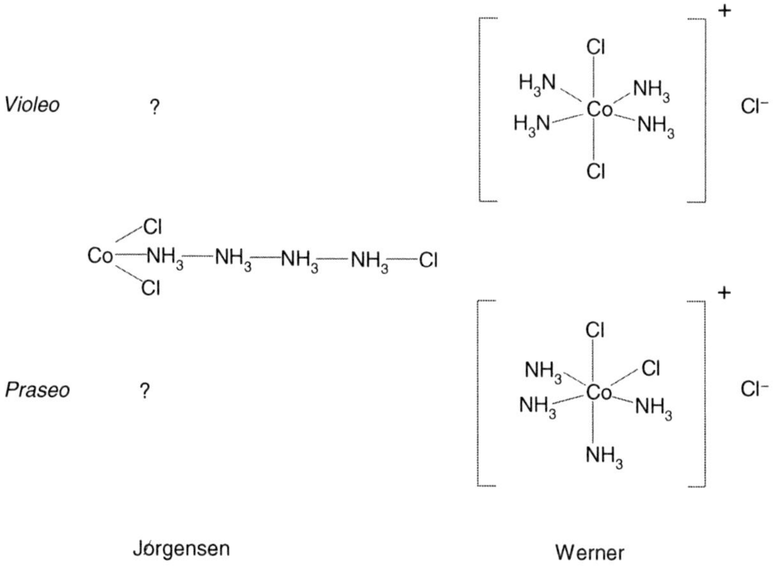

9.1 Misconceptions

Violeo ?

$$\left[\begin{array}{c} Cl \\ H_3N \diagdown \mid \diagup NH_3 \\ Co \\ H_3N \diagup \mid \diagdown NH_3 \\ Cl \end{array} \right]^+ Cl^-$$

Co—NH₃—NH₃—NH₃—NH₃—Cl (with Cl and Cl on the left Co)

Praseo ?

$$\left[\begin{array}{c} Cl \\ NH_3 \diagdown \mid \diagup Cl \\ Co \\ NH_3 \diagup \mid \diagdown NH_3 \\ NH_3 \end{array} \right]^+ Cl^-$$

Jörgensen   Werner

Fig. 9.5 Comparison of the chain theory and Werner's hypothesis [5]

verified his hypothesis with the synthesis of the two isomers (see Fig. 9.5). Frank Baeuerle and Mark Krasenbrink [6] present, in the German book on misconceptions [5], an excellent summary of coordination chemistry history and make good suggestions for teaching complex reactions in high school lessons [6].

9.1 Misconceptions

When looking through various textbooks, one can find about 10–20 pages of introduction regarding the concept of complexes, the examples given are mostly the well-known complexes with silver-, copper-, aluminum- and cobalt ions as central ions and H_2O- or NH_3 molecules or Cl^- ions as ligands. Inquiring in German high schools about the success of chemistry lectures concerning complex reactions, the answer from the chemistry teachers was that this topic is never taught: acid–base reactions and redox reactions are difficult enough. Even in articles or bibliographies [7], we found no references to misconceptions about complex reactions.

So, we decided to ask students in our University of Muenster using a special questionnaire. There are students studying chemistry with the aim of graduating with a Bachelor of Science in chemistry or a Bachelor of Chemical Education. Their aim is to become chemistry teachers. Both groups of about

60 students received different lectures and labs concerning complex chemistry. Martina Zwartscholten [8] and Christoph Lisowski [9], two chemical education students, working on their master thesis, developed the questionnaire and did the statistical work.

Questionnaire and results. In the studies of complex chemistry, there are three to four major subjects: composition and reactions of copper complexes, the stability of the tetraamminecopper complex, the equilibrium of aquacobalt(II) and chlorocobalt(II) complexes, and the solubility of silver chloride, silver bromide or aluminum hydroxide by forming the well-known soluble complexes. Using the ideas from the lectures in complex chemistry, the questionnaire was constructed and evaluated by a small group of students in their 3rd and 7th semester in college. Following this evaluation, the questions were rewritten and the students were invited to give answers. The answers were not restricted to formulas or equations, but drawings of mental models were also requested. We will show six questions, the correct answers as well as highlighting the students' major misconceptions.

Problem 1. The first problem is only a warm-up-question dealing with Alfred Werner, the founder of coordination chemistry. His name is strongly linked to the definition of a complex because he discovered a hypothesis based on the spatial octahedral structure of cobalt complexes in 1892. Through synthesis of these complexes the hypothesis could be confirmed later.

Problem 1: *Which person is known as the founder of complex chemistry?*
a) Friedrich Konrad Beilstein
b) Alfred Werner
c) Hermann von Fehling
d) Friedrich Woehler

Only 30% of the students were familiar with Werner – they have heard names like Beilstein, Fehling or Woehler and also marked these names in about the same percentage.

Problem 2. Now, we want to give the students the formula of the sodium hexafluoroaluminate compound and to ask them definitions of some important ideas concerning complexes; the right answers are marked with an arrow.

Problem 2: *Looking at the formula of the complex compound, $Na_3[AlF_6]$, identify the ligand, central ion, coordination number and cation:*

a) *ligand* → F^-
b) *central ion* → Al^{3+}
c) *coordination number* → 6
d) *cation* → Na^+

Only half of the students chose the right answers. Most students estimated "3" or "9" as the coordination number because of the index 3 or the sum of both

9.1 Misconceptions

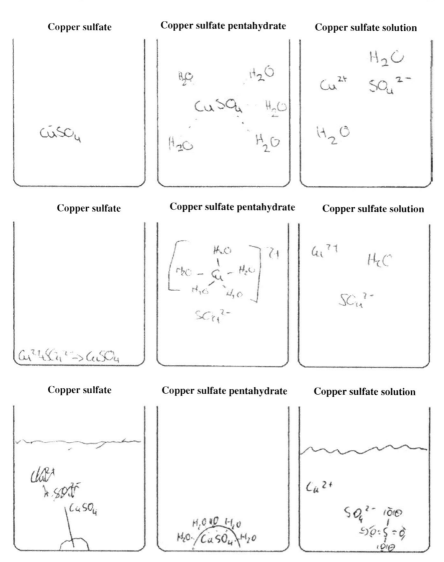

Fig. 9.6 Misconceptions of college students concerning copper sulfate complexes [9]

indexes 3 + 6 = 9. For some students, the central ion can be the whole complex "AlF_6" or the "cation, Na_3".

Problem 3. In this task, students have to show their mental model of well-known copper sulfate compounds by drawing symbols of the involved particles in given models of beakers. The diagram of Problem 3 represents one possible correct answer [9]. Only a few students could provide acceptable mental models for all three items, and more than two-third of the participants made severe mistakes: they exhibited many misconceptions (see Fig. 9.6).

Problem 3: Solid water free copper sulfate ($CuSO_4$) has a white color. If you add some water, dry blue copper sulfate pentahydrate ($CuSO_4$ 5 H_2O) is formed. If you dissolve it in water a blue solution appears. Draw symbols of existing particles in these beakers.

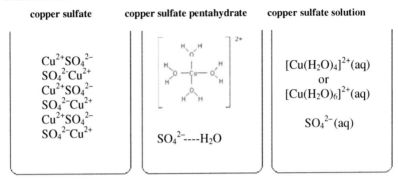

Some students' misconceptions (see Fig. 9.6) show that solid white copper sulfate is described without the involved ions. The name of copper sulfate *penta*hydrate makes many students think of the coordination number 5 – although in lectures they have all heard about the octahedral coordination of the copper ion (see Fig. 9.2). Some don't even take into account the coordinated copper ion but only of "$CuSO_4$" groups surrounded each by 5 H_2O molecules. The blue solution is mostly described by ion symbols – but only four students out of 60 have model images concerning complexes either with the coordination 6 (see Fig. 9.2) or with the coordination 4. Therefore, we will discuss the way in which the students should be taught and should learn coordination chemistry in their lectures (see Sect. 9.2).

Problem 4. A well-known experiment in the area of coordination chemistry is the reaction of cobalt chloride solution with hydrochloric acid: the bright pink color changes to deep blue; diluting this solution with water, the pink color returns. Scientifically, we think of the change from the aqua to the chloro complex, and of the equilibrium between both complexes:

$$[Co(H_2O)_6]^{2+}(aq, \text{ pink}) + 4\, Cl^-(aq) \rightleftharpoons [CoCl_4]^{2-}(aq, \text{ blue}) + 6\, H_2O$$

Problem 4: *If you add hydrochloric acid to a solution of cobalt chloride ($CoCl_2$), the color of the solution changes from pink to blue. If you dilute with water, the color changes from blue to pink. Please explain and show the chemical equilibrium.*

This reaction is part of the curriculum – but only very few students could provide an acceptable answer. Most students formulate wrong equations or answers, for example:

9.1 Misconceptions

- $2\,CoCl_2 + 2\,HCl \rightarrow 2\,CoCl_3 + H_2$
- $CoCl_2 + HCl \rightarrow CoCl_3 + H^+$
- $CoCl_3 + H_2O \rightarrow CoCl_2 + HCl + OH^-$
- if you dilute with water, the concentrated hydrochloric acid reacts and HCl will be protonated: $HCl + H_2O \rightleftharpoons H_3O^+ + Cl^-$.

There is some casual knowledge of hydrochloric acid and the protonation of "HCl" – but the idea of aqua- or chloro-complexes could not be realized by most of the students; there are nearly no mental models of the complexes learned in university lectures. In addition, the answers concerning equilibrium are not even mentioned (see also Chap. 6).

Problem 5. Precipitation reactions and the solubility of these precipitates by forming a soluble complex is an important aspect of coordination chemistry. One example is the precipitation of silver chloride using silver nitrate solution and a drop of concentrated hydrochloric acid and the reaction of silver chloride with an excess of the same acid. Possible right answers are given in the diagram of Problem 5 [9].

Problem 5: *If you take silver nitrate solution $(AgNO_3)$ and add a drop of hydrochloric acid (HCl), a white precipitate of silver chloride $(AgCl)$ is formed.*
a) Write, using an equation, those particles which are reacting:

$$Ag^+(aq) + NO_3^-(aq) + H^+(aq) + Cl^-(aq) \rightarrow Ag^+Cl^-(s) + H^+(aq) + NO_3^-(aq)$$

or:
$$Ag^+(aq) + Cl^-(aq) \rightarrow Ag^+Cl^-(s)$$

b) Show your mental model by drawing symbols of involved particles:

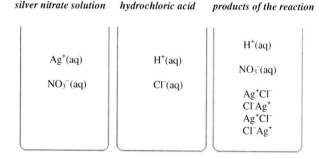

c) If you add more hydrochloric acid to the precipitated silver chloride $(AgCl)$, the solid dissolves and a colorless solution results. Write the chemical equation:
$$Ag^+Cl^-(s) + Cl^-(aq) \rightleftharpoons [AgCl_2]^-(aq)$$

d) *Draw your mental model of particles in the produced solution:*

> [AgCl$_2$]$^-$(aq)   H$^+$(aq)
>
>   H$^+$(aq)   [AgCl$_2$]$^-$(aq)
>
> [AgCl$_2$]$^-$(aq)   H$^+$(aq)

The precipitation of silver chloride is described by 60% of the students: the involved ions have been used correctly (see Problem 5a). Other answers involve formulas, but no ion symbols (see Fig. 9.7): this is not wrong but the question regarding the *reacting particles* is not answered (see Chap. 5). In Problem 5b, the model images of solutions before the reaction are correctly answered by about 80% of the students. Making model drawings of solid silver chloride presented the students with more difficulties (see Fig. 9.7).

The big difficulties arise with complex reactions of the solid silver chloride (see Problem 5c and 5d): only about 10% gave correct answers and drew acceptable models. The other students did not answer at all or offered incorrect equations and drawings – even knowing that coordination chemistry would be asked in the questionnaire. Most students offered equations like these:

- AgCl + HCl → Ag$^+$ + 2 Cl$^-$ + H$^+$
- AgCl + HCl + HNO$_3$ → Ag$^+$NO$_3^-$ + 2 HCl
- 2 AgCl + 2 HCl → 2 AgCl$_2$ + H$_2$.

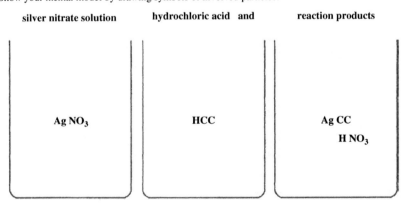

Fig. 9.7 Example for incorrect mental models about the precipitation of silver chloride [9]

9.1 Misconceptions

Problem 6. The production of aluminum is a big topic in chemistry curricula of universities and high schools – ending with the electrolysis of molten aluminum oxide. The first step in the whole process is the separation of aluminum hydroxide from the mineral bauxite, which is found in mines. Therefore, we asked the students about this separation because a complex reaction of aluminum hydroxide with sodium hydroxide solution is involved and students are familiar with this easily demonstrated experiment from the laboratory. Examples of correct answers are shown (see Problem 6a and 6b).

Problem 6: *The first step of aluminum production is the separation of aluminum hydroxide (Al(OH)₃) from the bauxite using concentrated sodium hydroxide solution (NaOH).*
a) *Show the chemical equation for dissolving solid aluminium hydroxide:*

$$Al(OH)_3(s) + OH^-(aq) \rightarrow [Al(OH)_4]^-(aq)$$

b) *Write down chemical symbols of the involved particles:*

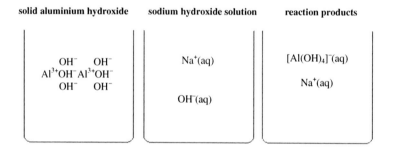

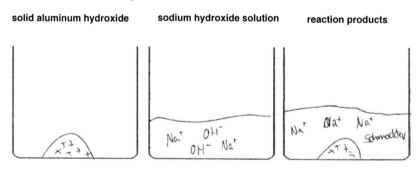

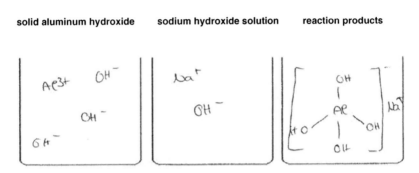

Fig. 9.8 Examples of mental models concerning the aluminum hydroxide reaction [9]

Because it was the last question of the questionnaire, half of the students did not even work on this problem; the other part made many mistakes with chemical reactions or with mental models. Only 15% of the students solved this problem completely. Two examples show some of the mistakes (see Fig. 9.8). Even though students knew the formula aluminum hydroxide, some had no mental model of an ionic lattice of aluminum ions or hydroxide ions (see Fig. 9.8). The reaction products are shown with symbols of sodium ions, but the student has no idea about the coordination compound. Another student had the right idea about the tetrahydroxo complex anion and even drew the sodium cation (see Fig. 9.8).

Result. The main result of the empirical research was as follows: knowledge concerning complex reactions seems very poor, very few memorized formulas remain from university lectures, but no mental models concerning complex particles. The students saw the bracket symbols like $[Cu(H_2O)_4]^{2+}$ in their lectures, but they got no real idea about the existence of these complexes as particles in solid blue copper sulfate crystals or in the solution of this salt. Students saw the complex formulas of the process of dissolving solid aluminum hydroxide by the well-known complex reaction – but could not understand these formulas.

It seems that writing equations is not enough in chemical education: the mentioned jump from the "macro level" to the "representational level" is too difficult (see Fig. 2.1). If we would introduce the "sub-micro level" by model drawings of the involved particles, students may develop better mental models in their cognitive structure and would more easily remember involved complex particles. We will try to make some teaching suggestions in this direction.

9.2 Teaching and Learning Suggestions

As in the area of acids and bases (see Chap. 7) and redox reactions (see Chap. 8), we would introduce complex reactions by the use of convincing experiments and good 2D or 3D model drawings. Therefore, the main goal for the introduction of coordination chemistry should be teaching the implicated complex particles.

Existence and Ligand Transfer. One experiment to convince students about complex particles is the following (see E9.1). Three different colored copper salts like white sulfate, green chloride and black bromide are dissolved in water and observed: all three solutions have the same bright blue color; typical in diluted solutions of copper salts. This color can be explained by the presence of the same kind of particles in all three solutions: the $[Cu(H_2O)_6]^{2+}$(aq) ions (see Fig. 9.2).

To convince students that the symbol Cu^{2+}(aq), by itself, is not enough to describe the situation, another complex particle should be introduced: the $[Cu(NH_3)_4]^{2+}$(aq) ion or the $[Cu(NH_3)_4(H_2O)_2]^{2+}$(aq) ion, that produces a deep violet solution. If one adds concentrated ammonia to all three blue-colored solutions, the color changes to deep violet in every solution (see E9.1). In all three cases, the following reaction takes place:

$$[Cu(H_2O)_6]^{2+}(aq, blue) + 4\ NH_3(aq) \rightarrow [Cu(NH_3)_4(H_2O)_2]^{2+}(aq, violet) + 4\ H_2O(aq)$$

So, the Cu^{2+}(aq) ion, alone, cannot explain the blue and violet color: there must be different particles, the aqua complex and the ammine complex. These complexes even exist in solids. If one evaporates the water from the blue solution, crystals of the same bright blue color remain:

$$[Cu(H_2O)_6]^{2+}(aq, blue) + SO_4^{2-}(aq) \rightarrow [Cu(H_2O)_4]^{2+} \cdot H_2O \cdot SO_4^{2-}(s, blue)$$

The structure of solid blue copper sulfate can be explained by $[Cu(H_2O)_4]^{2+}$ ions surrounded by sulfate ions in the way that O atoms of two sulfate anions form an octahedron with four water molecules (see Fig. 9.2). The "fifth" water molecule of "CuSO$_4$ 5 H$_2$O" is located between the aqua copper complexes (see Fig. 9.3).

The important explanation is still that these substances contain the involved complex ions plus water molecules in a 3-dimensional ionic lattice. If one continues to heat the blue crystal the color of the crystals turns white because the water molecules leave the ionic lattices and only the copper ions and sulfate ions are building up the new ionic lattice (see E9.2):

$[Cu(H_2O)_4]^{2+} H_2O\ SO_4^{2-}(s, blue) \rightarrow Cu^{2+}SO_4^{2-}(s, white) + 5\ H_2O$

It is helpful to show 3-dimensional models of theses complex particles. On the one hand the students could build an octahedron with cardboard and could demonstrate the meaning of each of the six corners (see E9.3), they should also compare that model with 2-dimensional drawings (see Figs. 9.2 and 9.3). The teacher could take a large Styrofoam ball and could insert six magnets at six corners of an octahedron for fixing models of H_2O molecules or NH_3 molecules (see E9.3). The ligand transfer could be shown with this model i.e. from aqua complexes to ammine complexes (see Fig. 9.9).

Coordination Number. In order to test for the appropriate coordination number of complex compounds, one could use the quite popular color effect, which is often used in complex chemistry. The addition of ligands or the exchange of ligands is associated with a shift in color, or even with a color change. For this reason, it is possible to visually examine the addition and the exchange of ligands. For instance, a green-colored mixture of hydrochloric acid plus copper sulfate solution is added to 10 test tubes (see E9.4). The first test tube is used as a blind test, 0.5 ml ammonia solution is added drop wise into the second test tube; 1.0 ml is added to the third test tube and with each following test tube, the volume is increased by 0.5 ml. A color change from green to deep violet is observed, the deep violet color remains in test tubes 6–10 (see E9.4). It can be assumed that the central copper ion is not capable of binding an unlimited number of NH_3 molecules.

A quantitative examination of the same phenomenon can be taken further by using a photometer [1]. Equimolar solutions of copper sulfate and ammonia are mixed together with this the amounts of NH_3 molecules are increased to a ratio of 1:7, measuring the amounts of involved complex ions (see E9.5). The extinctions of the solutions reach a maximum at a ratio of 1:4; this maximum remains constant for all higher concentrations of ammonia (see E9.5). With this, the

Fig. 9.9 Octahedron models of aqua copper and ammine copper complexes

9.2 Teaching and Learning Suggestions

coordination number 4 shows up for the copper complex, it is described as tetraamminecopper(II) ion or as a tetraamminediaqua copper(II) ion:

$$[Cu(H_2O)_6]^{2+}(aq, light\ blue) + 4\ NH_3(aq) \rightleftharpoons [Cu(NH_3)_4]^{2+}(aq, violet) + 6\ H_2O$$

$$[Cu(H_2O)_6]^{2+}(aq, light\ blue) + 4\ NH_3(aq) \rightleftharpoons [Cu(NH_3)_4(H_2O)_2]^{2+}(aq, violet) + 4\ H_2O$$

Model drawings of these solutions are ideal for demonstrating to students that the complex particles are free-moving ions, which are compensated by sulfate ions or chloride ions (see Fig. 9.10).

Complex Stability. In tetraamminecopper(II) ions, the central copper ion is surrounded by NH_3 molecules in such a way that it no longer shows usual reactions of free copper ions: it is assumed that copper is no longer deposited by an iron nail from the complex solution and that insoluble copper hydroxide no longer precipitates.

It is observed in the experiment that the iron nail immediately creates a copper deposit in a blue colored copper sulfate solution (see E8.1), whereby this does not happen in the violet colored ammine complex solution. A trace of copper deposit can only be observed after it has been dipped into the complex solution for a while (see E9.6). It is possible to verify this hypothesis with the help of a second reaction, the metal hydroxide precipitation (see E9.6): a greenish blue deposit is commonly observed in the blue solution of hexaaquacopper ions, but not in the solution of tetraamminecopper ions. Apparently, copper ions and water molecules are not very tightly bonded in aqua complexes, but copper ions and ammonia molecules in ammine complexes are: there is a weak stability of aquacopper ions, but a great stability of tetraamminecopper complexes. The stability constants can be taken and interpreted if one wants a quantitative explanation of these phenomena.

Complex Structure. The violet-colored solution of very stable nickel ammine complexes also confirms, through the described experiments, that practically no free nickel ions exist in the complex solution (see E9.7). The analysis of

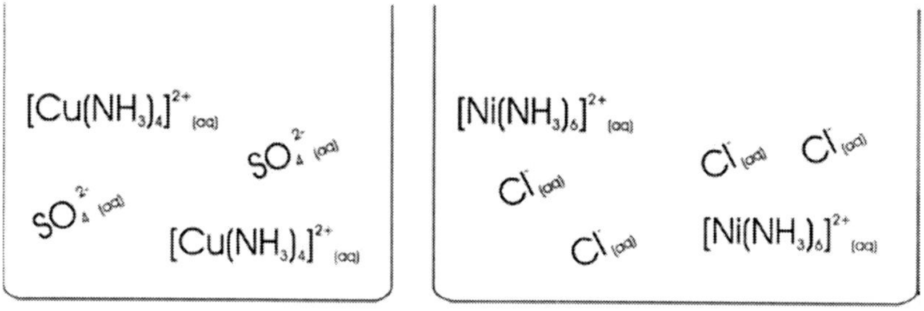

Fig. 9.10 Mental model for complex ions in aqueous solution

the coordinating number leads to the number 6; the complex is known as hexamminenickel(II) ion, [Ni(NH$_3$)$_6$]$^{2+}$(aq). A schematic model drawing could be used to demonstrate these complex ions (see Fig. 9.10). To have the mental model of the spatial structure of these complexes, a spatial model is useful (see Fig. 9.11): six NH$_3$ molecules surround the nickel ion octahedral; their N atoms are positioned on the corners of a regular octahedron.

Many complexes originating from cobalt ions comply with the regular octahedron structure. If one prepares the yellow crystalline hexamminecobalt(III) chloride from pink-colored cobalt(II) chloride, one obtains the octahedral structure of a hexamminecobalt(III) ion – it is a building unit of the crystal lattice. If it is dissolved in water, the complex ion remains intact:

$$[Co(NH_3)_6]Cl_3(s, yellow) - (aq) \rightarrow [Co(NH_3)_6]^{3+}(aq, yellow) + 3\ Cl^-(aq)$$

If one prepares the dichlorotetraamminecobalt(III) complex, one obtains either a violet or a green substance: there are two isomers of this structure (see Fig. 9.11). Both chloride ligands from the complex are to be found side-by-side in the cis-form or diametrically opposed in the trans-form. This example should – analogous to the meaning of isomers in Organic Chemistry – highlight the structure-property connections in complex ions. Many other examples and structures can be found in chemistry textbooks.

One could construct an octahedron using cardboard in order to demonstrate such an octahedral structure (see E9.3) with the corners representing the positions of the N atoms of six NH$_3$ molecules. In order, however, to create the central particle, one could use one large Styrofoam ball and six magnets in octahedrically-arranged positions of the surface (see E9.3): molecular models of NH$_3$ molecules could each attach thereon, if the model is magnetic for the N atom (see Fig. 9.9). One could take, for instance, a black spherical model from the Phywe molecular building set for N atoms: the press-buttons of these models are magnetic.

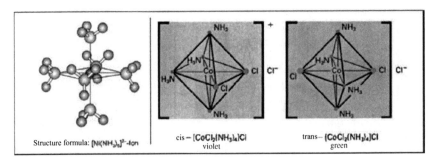

Fig. 9.11 Structure of a nickel complex [1] and isomers of cobalt complexes [10]

9.2 Teaching and Learning Suggestions

Complex Equilibria. If one dilutes green-colored copper chloride solution with water, it turns light blue. If, however, concentrated hydrochloric acid is added, the green color is regenerated (see E9.9) – a chemical equilibrium exists:

$$[Cu(H_2O)_6]^{2+}(aq, \text{light blue}) + Cl^-(aq) \rightleftarrows [CuCl(H_2O)_5]^+(aq, \text{green}) + H_2O$$

The structure of both complex ions correlates again to the octahedral structure (see Fig. 9.9). The stability of the hexaaquacopper(II) ion is so small that another ligand, for instance the chloride ion, readily replaces it (see E9.9).

If ammonia is added to the light blue copper sulfate solution, then the color turns deep violet: a ligand exchange takes place. However, the resulting complex is so stable that the equilibrium lies completely on the side of the ammine complexes. The color of the solution remains unchanged during the dilution with water; the ammine complex is not destroyed during the process:

$$[Cu(H_2O)_6]^{2+}(aq, \text{light blue}) + 4\, NH_3(aq) \rightleftarrows [Cu(NH_3)_4(H_2O)_2]^{2+}(aq, \text{violet}) + 4\, H_2O$$

The equilibrium of the following two cobalt complexes shows a change of the structure and of the coordination number. This equilibrium can also be demonstrated as being dependent on temperature. If hydrochloric acid is added to a pink-colored cobalt(II) chloride solution, the solution changes to blue due to the chloro complex (see E9.10). If the solution is diluted again with water, the pink color returns:

$$[Co(H_2O)_6]^{2+}(aq, \text{pink}) + 4\, Cl^-(aq) \rightleftarrows [CoCl_4]^{2-}(aq, \text{blue}) + 6\, H_2O$$

Adding solid sodium chloride to the pink-colored solution, the color does not change to blue but stays pink. If, however, the solution is warmed up, the color changes to blue (see E9.10): the reaction to the chloride complex is endothermic; the equilibrium favors the formation of this complex. After the solution has cooled down, the pink color returns (see E9.10).

The aqua complex changes from the octahedral structure with coordination number 6 to the chloro complex during the reaction. This shows a tetrahedral structure and exhibits the coordination number 4. A structure or coordination number is not to be seen as a permanent constant for a metal cation. Depending on size, charge and required space of the ligand, the coordination number can change; it is usually 2, 4 or 6. Further data can be taken from appropriate specific literature.

Even the chelate complexes offer many possibilities to demonstrate and to discuss related complex structures and equilibria. At any rate, it is possible to teach complex compounds used everyday and in industry. Detailed examples can be taken from textbooks.

Dissolution of Solids by Forming Complexes. If one takes an aluminum chloride solution and adds sodium hydroxide solution, two observations can be made. Using only a little amount of hydroxide solution, a precipitation occurs (see E9.11):

$$Al^{3+}(aq) + 3\ OH^-(aq) \rightarrow Al(OH)_3(s, white)$$

A large amount of hydroxide solution produces a colorless solution or the fallen precipitation of aluminum hydroxide is dissolved and the tetrahydroxoaluminum complex is formed (see E9.11):

$$Al^{3+}(aq) + 4\ OH^-(aq) \rightarrow [Al(OH)_4]^-(aq)$$

$$Al(OH)_3(s, white) + OH^-(aq) \rightarrow [Al(OH)_4]^-(aq)$$

These complex reactions are important for producing aluminum in industry. In the first step, one mixes Bauxite mineral with concentrated sodium hydroxide solution: the tetrahydroxoaluminum complex will remove all the aluminum of the Bauxite. Then, aluminum hydroxide must be changed to aluminum oxide; this compound is used in smelting flux electrolysis to finally produce pure aluminum metal.

There is also the other important process of producing black-and-white photos in the developing bath and in the fixing bath. The dissolution of unexposed silver bromide on the photographic paper in the fixing bath can be described by a complex reaction (see E9.12):

$$AgBr(s, yellow) + 2\ S_2O_3^{2-}(aq) \rightarrow [Ag(S_2O_3)_2]^{3-}(aq) + Br^-(aq)$$

It might be interesting for students to see the development of photos in the laboratory and to know the theory behind it. Many other applications involve complex reactions that one could explore and explain.

Historic Approach. In order to offer students the chance to make their own independent inquiries on the structure of complexes, one can look at the historically problem-oriented approach [11]: the students carry out a series of experiments on complex compounds and discuss the historical literature on different theories of the past. The students could, for instance, become acquainted with the discussions lead by the chemists of that time, Jorgenson and Werner. Students can be asked to summarize the positions of both chemists or debate, in small groups, the position of either or both of these chemists, thereby developing a deeper understanding of the background related to the chemical and historical facts. The following texts would be suitable for use in the group work [6]:

9.2 Teaching and Learning Suggestions

Literature on Modern Complex Theory.

- Gade, L.H. : Alfred Werners Koordinationstheorie, ChiuZ. 36 (2002), 168–175
- Wannagat, U.: Das Portrait: Alfred Werner, ChiuZ. 36 (2002), 24–27
- Bailar, J.C.: The Chemistry of Coordination Compounds, New York 1956
- Kauffman, G.B.: Classics in Coordination Chemistry – The Selected Papers of Alfred Werner, New York 1968

Literature on the Historic Chain Theory.

- Werner, A.: S. M. Jorgenson 1837–1914, Chem. Z. 36 (1914), 557–564
- Mäueler, G.: Zum Wandel von Theorie, Nomenklatur und Formelsprache der Koordinationsverbindungen, PdN-Ch. 33 (1984), 103–111
- Bailar, J.C.: The Chemistry of Coordination Compounds, New York 1956
- Kauffman, G.B.: Classics in Coordination Chemistry – Selected Papers, New York 1976

A fictitious Chemistry Congress can be simulated as a game, in which the followers of the theories of Jorgenson and Werner can meet up and discuss their ideas. At first, each group should debate, using the relevant model concepts, and should then do their best to prove that the opposing team's model is not sufficient to appropriately explain the observations recorded in the experiments. In addition, it should be shown by both groups, why the model used is exceptionally appropriate for explaining the observed phenomena. The Werner representatives should be able to overthrow Jorgenson's followers in the debate.

A great advantage of this method is that the students practice how to critically examine models: they learn that models are helpful in explaining phenomena, but can never be a true representation of reality. This series two purposes: firstly, it's possible to use several models to examine one and the same phenomenon, without being able to determine that one model is better than the other. Secondly, students recognize quite clearly, that scientific mental models are continually being changed through new experiences, making it almost impossible to see one particular model as being the "correct one" for all time.

Furthermore, one has the rare opportunity of comprehending this phenomenon through the use of additionally gained information, from the initial mental model up until the concrete demonstration model. This is why one should use the opportunity of setting up these lessons as a structure-oriented approach [12]. The first path would be the observations of the laboratory experiences and phenomena, whereas the second path would be the use of scientific explanations of these phenomena, based on the kinds of particles, structural models, like 3D-octahedrons, and model drawings, and finally the development of related mental models [12].

Outlook. Are there complex reactions suitable for use in school lessons? Alternatively, is one in danger of becoming entangled in abstract scientific theories, which are far too theoretical? The surveys of teachers at Muenster high schools show that this topic is not at all taught in chemistry classes. However, it is merely speculative as to whether this is because there is just too much information to cover, or due to its complexity.

Looking at the amount of material on this topic should lead to a useful didactical reduction: if only the shown complexes of copper, silver and aluminum are offered to students, they have the chance for successful learning. The complexity can also be reduced with the use of model material like 3D-octahedron and model drawing, and by developing mental models of these complex particles. If acid–base reactions and redox reactions are taught to students, the basic idea of equilibria can be taken to develop the idea of complex reactions. Finally, convincing experiments can be demonstrated or done by students themselves, as will be shown below.

9.3 Experiments on Complex Reactions

E9.1 Existence of Complex Ions

Problem: Students know the ionic symbols like Na^+(aq) and Cl^-(aq): with the (aq)-symbol one will show that about four to six water molecules are surrounding one ion attached by electrostatic forces. With a simple experiment about the dissolving process of three different copper salts, the student will conclude that the Cu^{2+}(aq) is responsible for the same blue color of all three solutions. Adding ammonia solution to all three liquids, an identical deep-violet solution results. What particles are now responsible for this new color? The teacher has to introduce copper complexes, symbolized by special symbols $[Cu(H_2O)_6]^{2+}$ and $[Cu(NH_3)_4(H_2O)_2]^{2+}$. These six ligands in both complexes combine with the central ion by a special structure which can be shown by octahedron models (see Figs. 9.2 and 9.9).

Material: Test tubes; white copper sulfate, green copper chloride, black copper bromide, water, concentrated ammonia solution.

Procedure: To each of the three copper salts, add a little amount of water; shake the test tubes. In a second step, add a few mls of ammonia solution to all three solutions.

Observation: Although the salts initially have different colors, all the three solutions are bright blue. They turn into the same deep violet color after adding ammonia solution.

E9.2 Blue and White Copper Sulfate

Problem: Because the light blue color of the solid copper sulfate hydrate is identical to the aqueous solution, students should be convinced that the

9.3 Experiments on Complex Reactions

same aqua complex is present in both substances. The complex can be destroyed in solution through ligand transfer, for instance with ammonia molecules: the color changes to deep violet. It is also possible to destroy the complex by heating the solid crystals: copper sulfate hydrate reacts to water and anhydrous copper sulfate; in this case the color changes from light blue to white. To regenerate the blue crystals, water is added to the white copper sulfate.

Material: Test tubes, burner, copper sulfate hydrate ($CuSO_4$ 5 H_2O).

Procedure: Fill a test tube with several spatula tips of copper sulfate hydrate and clamp the test tube slanted pointing downwards. Heat the salt intensely. Cool the white salt and then add several drops of water.

Observation: A white substance remains from the blue copper salt; water forms during the heating process and drips out of the test tube. When water is added to the white substance, light blue copper salt returns, the contents of the glass become quite hot.

Tip: If portions of the substances are quantitatively weighed before and after the experiment, one achieves a good result, which shows the contents of 5 mol water molecules per mol copper sulfate hydrate, thereby establishing the existence of copper sulfate pentahydrate.

E9.3 Octahedral Models of Complex Particles

Problem: Structural models are useful in order to raise the clarity in the discussion of complex structures and to develop mental models in the cognitive structure of students. Students should build the models first; then, they could draw them in perspective. In addition, simple cardboard octahedrons or tetrahedrons could be used. Structural models could also demonstrate the central particle with the help of molecular building sets. The octahedron, especially, serves the purpose of demonstrating the octahedral structure of $[Cu(H_2O)_6]^{2+}$ and $[Ni(NH_3)_6]^{2+}$ (see E9.7), respectively $[Co(NH_3)_6]^{2+}$ (see E9.8).

Material: Cardboard and glue, Styrofoam ball (d = 10 cm) and Styrofoam glue, 6 small magnets, molecular building set (similar to the Phywe System).

Procedure: (a) Enlarge nets on the building of octahedrons and tetrahedrons (picture) to such an extent that the edge length is about 10 cm on the cardboard. Cut each net and glue together. (b) Cut holes at six octahedral positions in the large styrofoam sphere, and glue small magnets in these holes. Construct six models for H_2O molecules and NH_3 molecules out of black and white spheres of the Phywe-molecular building set; and attach them to the magnets of the styrofoam sphere (see Fig. 9.9).

Observation: The models show complex structures of the coordination numbers 4 (tetrahedron) and 6 (octahedron).

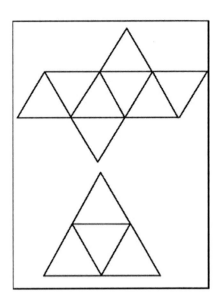

E9.4 Properties of the Ammine Copper Complexes

Problem: The coloring of various complex compounds is suitable to experimentally show, that a specific rather than an arbitrary amount of ligands per central ion is bonded. For instance, through the successive additions of ammonia solution to copper sulfate solution the deep violet colored solution of the tetra ammine copper complex is finally achieved. With this, it is possible to demonstrate that a specific amount of ammonia solution is necessary to get the violet colored ammine complex, any further additions do not lead to other changes in color.

Material: Beakers, test tubes, plastic pipettes, measuring cylinders (100 ml), scales; copper sulfate hydrate, 2-M hydrochloric acid, ammonia solution (25% solution, diluted 1:50).

Procedure: Weigh 1 g copper sulfate, add 40 ml hydrochloric acid and dissolve. Label 10 test tubes from 1 to 10 and fill each with 3 ml of this green colored solution. Use test tube 1 as reference. Add ammonia solution to test tubes 2–10. Starting with the addition of 0.5 ml in test tube 2, with 1.0 ml to test tube 3, increase the volume of ammonia solution by 0.5 ml per test tube. Finally, by adding water, make the volume of all test tubes equal.

Observation: A gradual color change from green to dark violet can be seen from test tubes 2 to 5. The intensity of the violet color remains constant from test tube 5 until test tube 10.

E9.5 Determination of the Coordination Number

Problem: The previous experiment roughly shows, qualitatively, that there is a specific amount of NH_3 molecules, which is attached to the central copper ion in

9.3 Experiments on Complex Reactions

ammine complexes (see E9.4). If a similar experiment is carried out with equimolar solutions and if a photometer is used for color absorption, the coordination number 4 can be quantitatively achieved: with a ratio 1:4 the extinction should be maximal, regardless of the amount of NH_3 molecules available.

Material: 8 small beakers, plastic pipettes, measuring cylinders, scales, photometer; 0.1-M solutions of copper sulfate and ammonia, ammonium nitrate.

Procedure: Prepare mixtures A – H as indicated below. In order to avoid the precipitation of copper hydroxide during mixing, add 10 g of ammonium nitrate beforehand and dissolve. Measure the extinction of all solutions at a wavelength of 600 nm using the photometer.

	A	B	C	D	E	F	G	H
V(CuSO$_4$) in ml	10	10	10	10	**10**	10	10	10
V(NH$_3$) in ml	–	10	20	30	**40**	50	60	70
V(water) in ml	70	60	50	40	**30**	20	10	–
Extinctions:	0.01	0.07	0.25	0.44	**0.54**	0.56	0.57	0.58

Observation: Extinction measurement values range from 0.01 to 0.58: the maximum extinction is reached in solution E, after that the extinction remains practically constant.

E9.6 Copper Complexes and Stability

Problem: In order to show further phenomena on the structure of complexes and complex equilibria, it should be shown that the central ion is solidly bound to the ligands and is not solely present in the solution, in the stable tetra ammine copper complex. In order to do this, an iron nail is dipped into the complex solution, respectively, diluted sodium hydroxide solution is added and this is compared to regular copper sulfate solution: the iron nail does not show the copper deposit as usual, no precipitation of the copper hydroxide is deposited. The copper sulfate solution should be interpreted in comparison to the complex solution as a solution with free Cu^{2+}(aq) ions or very instable aqueous copper complexes. With the explanation of the copper deposit on iron a cross-linkage to redox reactions (see Chap. 8) is possible.

Material: Test tubes, sand paper; iron nail, copper sulfate hydrate, diluted ammonia solution, diluted sodium hydroxide solution.

Procedure: Put diluted blue copper sulfate solution into four test tubes. Add ammonia solution to two test tubes until the deep-violet color appears, don't add ammonia to the other two test tubes. Have two sets of solutions one with and one without ammonia. Add a freshly sand-papered iron nail to the first set of solutions; to the other set a little amount of sodium hydroxide solution.

Observation: A copper deposit forms immediately in the light blue solution; this can easily be seen when the nail is removed from the solution. A copper deposit does not form immediately in the deep violet solution (only after quite some time). In the light blue solution greenish blue copper hydroxide is deposited, but not in the deep violet solution.

E9.7 Nickel Complexes and the Octahedral Structure

Problem: The copper aqua complexes which have been discussed thus far cannot be clearly described [13]: there are four H_2O-ligands arranged around the central copper ion in planar formation, two further ligands at a somewhat larger distance; their O atoms, together with those of the first mentioned four ligands form the corners of a distorted octahedron. This is why the copper aqua complex is normally described with the coordination number 4, in other places with the coordination number 6 [13]. In order to interpret a complex with a clear and regular octahedral structure (see Fig. 9.11) and to construct corresponding octahedral structural models (see E9.3), the hexa ammine nickel complex $[Ni(NH_3)_6]^{2+}$ (aq) is demonstrated. The stability of nickel aqua complexes is very low, of nickel ammine complexes very high: no solid nickel hydroxide will precipitate.

Material: Test tubes; solution of nickel chloride hydrate $[Ni(H_2O)_6]Cl_2$, diluted ammonia solution, diluted sodium hydroxide solution.

Procedure: Gradually add, in portions, ammonia solution to the green-colored nickel chloride solution until the color turns violet. Show and compare a 3D-model for the octahedron structure of the hexa ammine complex. Slowly add sodium hydroxide solution (drop by drop) into the violet-colored solution, for comparison also to the green-colored nickel chloride solution.

Observation: The color of the solution changes from green, through several other colors, to violet. No precipitation appears from the violet solution after adding sodium hydroxide solution, this does however happen in the green solution.

E9.8 Synthesis of Hexamminecobalt Chloride Crystals

Problem: In order to also obtain a solid substance which shows a regular octahedral complex, from pink colored cobalt(II) chloride hydrate the yellow colored ammine complex of cobalt(III) ions is formed: $[Co(NH_3)_6]^{3+}$. However, the cobalt(II) chloride has to be oxidized with hydrogen peroxide – a cross-linkage to redox reactions (see Chap. 8) is possible.

9.3 Experiments on Complex Reactions

Material: Beaker (250 ml), round flask (100 ml), measuring cylinder (50 ml), scales, hot plate, Buechner-funnel, suction support, water jet pump; cobalt(II) chloride ($CoCl_2$ 6 H_2O), ammonium chloride (NH_4Cl), concentrated ammonia solution (25%), hydrogen peroxide solution (30%), active carbon granulates, ethanol.

Procedure: Dissolve 4 g of cobalt chloride and 3 g of ammonium chloride in 5 ml of water. Add 1 g active carbon and 20 ml ammonia solution to this solution, finally add slowly 5 ml hydrogen peroxide solution to this suspension. Heat the strongly frothing solution for 5 min. Filter the hot suspension through a Buechner-funnel which has previously been stored for several minutes in an oven at 100°C. Collect the filtrate in a round flask and place it in the refrigerator until yellow crystals are formed. Filter the crystals from the solution, wash in an ethanol-water mixture (2:1), and finally once more in pure ethanol.

Observation: Yellow crystals of the hexa ammine cobalt(III) chloride are formed.

E9.9 Copper Complexes and Equilibrium

Problem: The reactions on the transfer of ligands are always equilibria reactions – even in cases of stable ammine complexes the equilibrium lies far in the direction of the complexes. In order to demonstrate equilibria by experiments and to adjust them depending on concentrations, the equilibria between the less stable chloro- and aqua copper complexes are suitable.

Material: Test tubes; copper(II) chloride hydrate, solid sodium chloride, concentrated hydrochloric acid.

Procedure: Add water to two portions of green-colored copper chloride solution, then add solid sodium chloride, to the first portion, to the second one hydrochloric acid, and finally add water to both solutions.

Observation: The color of the solutions changes firstly from green to light blue, again to green, finally back to light blue.

E9.10 Cobalt Complexes and Equilibrium

Problem: In order to intensify the idea of equilibrium and the dependence on temperature, cobalt complexes are useful: with a mixture of water molecules and chloride ions as ligands for cobalt complexes the blue colored tetra chloro cobalt complex proves stable in heat, the pink colored hexa aqua cobalt complex in the cold solution, respectively. A cross-linkage to the idea of equilibrium (see Chap. 6) is possible.

Material: Test tubes, burner; cobalt(II) chloride solution, concentrated hydrochloric acid, sodium chloride.

Procedure: Add hydrochloric acid to the pink-colored cobalt chloride solution, dilute the obtained blue-colored solution with water. Put cobalt chloride solution once again in a second test tube; this time add several spatula tips of solid sodium chloride and dissolve it by shaking. Heat the solution with the burner, and cool the resulting blue solution in tap water. Repeat heating and cooling of the solution as often as desired.

Observation: The color of the solution changes from pink to blue when hydrochloric acid is added; when diluted with water the pink color returns. When the second solution is heated, the blue color appears; however, this changes again to pink when solution is cooled off.

E9.11 Dissolving Aluminum Hydroxide by Complex Reactions

Problem: In order to present students with a cognitive conflict it is possible to precipitate a solid and to dissolve this solid by the same substance. One way is to add a small amount of sodium hydroxide solution to aluminum chloride solution: aluminum hydroxide precipitates as a white solid. After that an excess of sodium hydroxide is added: the white solid disappears because the soluble tetrahydroxide aluminum complex is formed. If this excess of hydroxide solution is added from the beginning, no precipitation occurs, the complex is just formed.

Material: Test tubes; aluminum chloride ($AlCl_3$ 2 H_2O), sodium hydroxide solution.

Procedure: Fill a test tube to one-third with aluminum chloride solution, add some drops of sodium hydroxide solution. After observing a precipitate add some more sodium hydroxide solution slowly: the white solid dissolves. In a second test tube add to the aluminum chloride solution a large amount of sodium hydroxide at once.

Observation: In the first step a white solid is formed, in the second step the precipitation disappears. In the second test tube no precipitation occurs at all.

Tip: The association of industrially produced aluminum hydroxide from Bauxite can be discussed: aluminum hydroxide is separated from the accompanying substances of the Bauxite by mixing the pulverized mineral with hot concentrated sodium hydroxide solution.

E9.12 Dissolving Silver Halogenides by Complex Reactions

Problem: Because the process of black-and-white photography is well-known to students, they may want an explanation on how to fix a photo in the fixing bath. Like in the above experiment (see E9.11) a solid is dissolved by a complex compound, in this case silver bromide is dissolved by the dithiosulfatesilver complex: $[Ag(S_2O_3)_2]^{3-}$(aq). To prepare the students by using similar but simple reaction, silver chloride is firstly dissolved by an excess of hydrochloric acid: $[Ag(Cl)_2]^{-}$(aq) ions are formed.

Material: Test tubes; silver nitrate solution, concentrated hydrochloric acid, sodium bromide solution, concentrated sodium thio sulfate solution.

Procedure: Fill two test tubes to one-third with silver nitrate solution. In the first test tube add only one drop of hydrochloric acid, a few seconds later again add some ml of the acid until the white solid disappears. Add some drops of sodium bromide solution, to the second test tube a few seconds later add some ml of sodium thiosulfate solution.

Observation: In the first test tube a white solid precipitates: silver chloride. After adding more acid, the solid is dissolved to a colorless solution: dichloro silver complexes. In the second test tube a yellow precipitation occurs: silver bromide. After adding the sodium thiosulfate solution the solid dissolves and a colorless solution remains: dithiosulfate silver complexes.

Tip: To apply this knowledge in the photo laboratory the students may look to the process of making black-and-white photos, specially working with the fixing bath. It may be helpful to first have the experiences in the photo laboratory and to give all demonstrations and interpretations afterwards.

References

1. Asselborn, W., u.a.: Chemie heute, Sekundarbereich II. Hannover 2003 (Schroedel)
2. Wells, A.F.: Structural Inorganic Chemistry. Oxford 1987 (Clarendon Press)
3. Basolo, F.; Johnson, R.C.: Coordination Chemistry – The Chemistry of Metal Complexes. New York 1964
4. Werner, A.: Beitrag zur Konstitution anorganischer Verbindungen, Z.anorg. allgem. Chemie 3 (1893), 267
5. Barke, H.-D.: Chemiedidaktik. Diagnose und Korrektur von Schuelervorstellungen. Heidelberg 2006 (Springer)
6. Baeuerle, F., Krasenbrink, M.: Komplexchemie – Unterricht zwischen historischen und aktuellen Modellvorstellungen. Staatsexamensarbeit. Muenster 2005
7. Pfundt, H, Duit, R.: Students' Alternative Frameworks and Science Education. Kiel 1994 (IPN)
8. Zwartscholten, M.: Komplexchemie: Fehlvorstellungen bei Studierenden der Chemie. Staatsexamensarbeit. Muenster 2007
9. Lisowski, Ch.: Komplexchemie – Diagnose und Praevention von Fehlvorstellungen. Staatsexamensarbeit. Muenster 2007
10. Amann, W., u.a.: elemente chemie II. Stuttgart 1998 (Klett)
11. Matuschek, C., Jansen, W.: Chemieunterricht und Geschichte der Chemie. PdN-Chemie 34 (1985), 3
12. Barke, H.-D.: Strukturorientierter Chemieunterricht. In: Barke, H.-D., Harsch, G.: Chemiedidaktik Heute. Heidelberg 2001 (Springer)
13. Holleman, A.F., Wiberg, E.: Lehrbuch der Anorganischen Chemie. Berlin 1985 (de Gruyter)

Fig. 10.1 Concept cartoon concerning combustion of coal

Chapter 10
Energy

Energetic processes remain mysterious especially if one thinks of the manifold esoteric references and historical or biblical traditions, which have been passed on over generations. Scientists have unraveled the phenomenon of "energy" over the centuries and have described various types of energy, for instance potential and kinetic energy, heat energy, light energy, electrical energy, nuclear energy or chemical energy. However, much doubt remains as can be seen by Duit's statement regarding the teaching of the term Energy: "Is there an appropriate educational manner to present energy, a concept which, on the one hand is suitable for the experiences of students and, on the other hand, can be accepted by physics and chemistry" [1]. Feynman goes so far as to ascertain: "It is important to realize that in modern Physics, we do not know what energy really is" [1].

Nevertheless, we can describe and quantitatively measure energy. Historically, one means of measuring thermal energy is the **calorie** with the unit of 1 cal. What is meant by thermal energy of 1 cal is the amount of energy required to increase the temperature of 1 g water by 1°C (specially 14.5–15.5°C). Using the SI units (Système Internationale, 1967) like kg, m and s, the **joule** became an important energy unit: 1 cal = 4,18 J. It is derived from the term: $1 J = 1 \, kg \, m^2/s^2$, and can be interpreted as follows: energy that must be used in order to move an object weighing 1 kg with the constant acceleration of $1 \, m/s^2$ over a distance of 1 m.

Energy can neither be lost nor can it be gained. Generally, it is transformed from one form to another. For instance, when cycling with a running dynamo, the mechanical energy of the wheel is transformed into electrical energy through the dynamo. There are energy changes in a light bulb: electrical into heat and light. In a nuclear energy station, energy, which is obtained through nuclear fission of uranium, is transformed into thermal energy; this again is mechanically changed into electrical energy by the use of turbines, finally electrical energy is leaving the plant. This form of energy is ideal to transport over large distances, finally supplying each household with the required amount of energy: thermal energy for cooking, electrical energy for lighting, mechanical energy for cutting grass, etc. Many publications mention misconceptions related to energy and particularly concerning electrical energy: Pfundt and Duit [2] cite a large number of publications. In this chapter misconceptions relating to chemical energy will mainly be discussed.

Chemical energy and the transformation to other energy forms are important in chemistry lessons. However, this form of energy is not particularly graphic. Although we all transform energy-rich substances like sugar and starch in our bodies on a daily basis, to energy-poor ones like carbon dioxide and water, using the energy difference for heating our bodies, most of us are often not aware of this process. The opposite is true in nature: relatively energy-poor substances like water and carbon dioxide are transformed to sugar and starch (substances with a relatively high chemical energy value), through the addition of light and thermal energy. Burger and Gerhardt [3] stated, after conducting tests with "energy in biological context", that associations "are mainly correlated to the technical area. Most associations stemming from the area of animated nature... are mainly related to the areas of 'Man' (activities, muscles, body, nourishment) and 'Nature' (abiotic factors like climate occurrence, especially warmth, organisms like animals and plants)" [3]. These biological aspects are not discussed.

Young people do not really stand a chance of avoiding misconceptions or of attaining appropriate scientific concepts due to the lack of correct mental models and especially through their common everyday verbal exposure. As long as parents and friends speak of "empty batteries" or of "fuel consumption", are filling up at the gas (petrol) station and start off again with "new energy", neither the concept of the maintenance of energy nor the concept of transformation of one energy form in another can be developed. As long as "used energy" has to be paid to the energy companies or the "energy is used" in an electric light bulb and of the "loss of energy" in heating systems is discussed, it is not possible to establish scientific concepts in relation to energy.

10.1 Misconceptions

Many empirical research studies deal with "loss and gain of energy". Barker [4] found misconceptions by observing children looking at a methane flame: "energy is stored in methane; energy comes from burning the methane, from the flame, simply from the methane, fuels are energy stores". In other directions, she found that "energy is used up or lost; for all energies there is something that activates them; that gives the strength" [4]. She proposes in her "Implications for Teaching" [4] that "teaching energy as energy transfer is extremely important in developing the idea that in fact energy is not 'used up', but moves from one form to another".

Looking to "Children's Ideas in Science", Driver [5] found several other alternative conceptions taken from everyday life. Erickson and Tiberghien [6] show that children like "to talk about heat in terms of a 'state of hotness' of a body along a continuum from cold to warm to hot". Children tend to associate heat with living objects or with sources of heat: "heat rises up, the sun has it, the sun burns it and it shines and comes down and make the earth hot; heat moves from

10.1 Misconceptions

the sun to the air". Other pupils equated the idea of heat with a hot body or substance: "heat is warm air; heat is a warming fluid or solid; when you touch heat it feels hot".

Connecting the ideas of heat and temperature, they found: "Pupils thought that objects of different materials in the same room had different temperatures. For many pupils, metal objects were colder than wood objects. For example, a child who was asked if a container, full of water, was left for a long time in a room would be colder, hotter or the same as the water inside it, said: The container will be colder than the water" [6]. "In certain situations, many pupils appear to believe that the temperature of an object is related to its size. For example, they thought 'that a larger ice cube would have a colder temperature than a small ice cube' and hence the larger ice cube would take longer to melt; the larger objects contain more heat and therefore are likely to have a hotter temperature" [6].

Erickson and Tiberghien [6] show detailed research studies concerning pupils' conceptions about mixing of water. "Two basic types of situation are used: (1) similar amounts of water at the same temperature are mixed and (2) both similar and different amounts of water at different temperatures are mixed" (see Fig. 10.2). The results are shown in two diagrams (see Fig. 10.2): In a qualitative way, many pupils "at least acknowledge that the final temperature should lie somewhere between the initial temperatures", the quantitative calculations come later: "it is not before the age of 12 or 13 that a task like that in Fig. 4.2d is solved" [6].

Other researches show the connection between temperature and phase changes. Using questions about the boiling point of water, "the majority of 12–15 year-olds predicted that the temperature of boiling water remains at 100°C so long as the switch setting on the hot plate remained constant. If this setting was increased, then they predicted that the temperature of the boiling water would increase. It would seem that pupils could easily learn the fact that water has a boiling point of 100°C and that it may remain unchanged in certain conditions; however, most do not appear to have any clear understanding of why the temperature remains the same during a phase change. This understanding would seem to require some explanation at the molecular level..." [6].

"Pupils of all ages also experienced difficulty in differentiating between 'heat' and 'temperature'. Typical responses were that either temperature is 'a measurement of heat' or it is 'the effect of heat'. Some examples are: Temperature is the amount of heat, and heat raises the temperature; well temperature, it is just like a thing – like the sun – when you get the sun shining you get a temperature then. However, heat, you have to get something to make heat. But for temperature, it just comes, it's just natural temperature" [6].

Similarly, Kircher [7] show misconceptions about heat and temperature. He tells the story about a girl who investigates, if an ice cube wrapped in wool would melt quicker than another ice cube packed in aluminum foil. The girl thinks the ice cube wrapped in wool should melt quicker: "A wool shirt will keep me warm, the wool provides heat", are her arguments. By asking about the measured temperatures of a thermometer wrapped in wool and another one

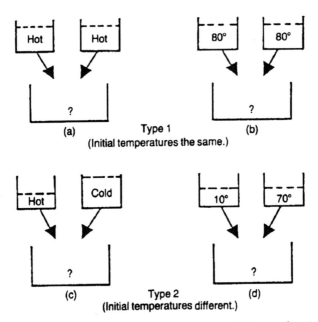

Figure 4.1: Four water-mixing questions requiring qualitative and quantitative responses.

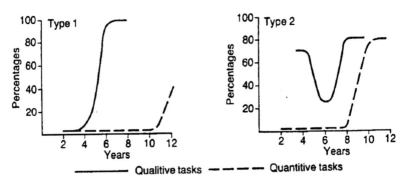

Figure 4.2: The success rates of pupils at different ages on the two types of water-mixing questions

Fig. 10.2 Questionnaire and results of conceptions concerning mixing of water [6]

lying nearby on the same table, a higher temperature is expected to be found in the one wrapped in wool. Even after doing the experiment, pupils are not convicted because they have ingrained experiences that a shirt of wool keeps them warm [7]. If you ask about the amount of energy which is necessary to heat a big cube of ice and a little one, many pupils think that the energy is the same: "the little one melts quicker of course" [7]. If you heat the same volumes of water and ethanol from 20 to 30°C, you need two minutes for ethanol, but four minutes for water. When asked to which liquid a higher energy is transferred, the answer is very often: "Both energies are the same because the same temperatures are reached" [7].

The reasons for the misconceptions relating to heat are based on two different everyday life experiences. On the one hand, people discuss the flow of heat from a hot object to a cold one: the word 'heat' is used to denote the direction of energy. On the other hand, people are thinking of temperatures of the upper part of the scale of the thermometer: the word 'heat' is used in the direction of a temperature [7].

Schmidt [8] and Griffiths [9] published some reflections related to "research on students' chemistry misconceptions". In the area of matter, Griffiths states: "Students at all levels appear to believe that the physical properties of atoms and molecules mirror those of the macroscopic substances involved. Thus, water molecules in ice are cubic in shape, molecules get hot or cold, as a substance is heated or cooled, and the particles of a substance are considered to melt when the substance is observed to melt" [9]. According to the idea of heat, he found statements like: "bubbles in boiling water are made of heat, bubbles are made of air, of oxygen and hydrogen; when water evaporates it ceases to exist, it changes to air; heat and light are forms of matter" [9].

10.2 Empirical Research

Everyday life conceptions about heat and temperature prevent the formation of acceptable mental models on energy. Grasping the idea of chemical energy stored in substances and being transferred by chemical reactions is even more difficult. Therefore, our own investigations should deliver more information on chemical energy. Tobias Doerfler [10] did these in German schools around Muenster. Using a questionnaire he asked about 200 students 15–18 years of age.

As a "Warm-up Question", the students were encouraged to discuss issues relating to the following: "Explain in a few sentences your concept of energy. Give an example". The majority of test persons quoted various definitions related to the term energy that appear to have been learned off-by-heart and most likely not really been understood: "energy is the ability to carry out work" or "energy is stored work". These sentences have clearly not been formulated by the students themselves, but rather stem from Physics lessons. This suspicion is confirmed by an answer supplied by one girl: she placed her formulated definition in quotation marks and mentioned that the teacher is the source of the information.

A 10th grade boy gave a further example: "energy is created where work is produced: kinetic energy, electrical energy". This pupil clearly depicted his concept of energy in a very concrete way: using the example of riding a bicycle, he believes that energy "is created" and misjudges the fact that his movement during cycling also creates a form of energy, i.e. mechanical energy, which is transformed to a different form of energy, i.e. electrical energy.

Other examples naturally stem from one's clearly depicted experiences. For example, electrical energy is associated with electricity, light bulb or lightning, kinetic energy with explosions, thermal energy with the burning of fuel. A 9th grade pupil writes: "energy for me is when a car explodes or a balloon bursts!" Others state: "energy is created when fire contacts fuel" or "energy can be produced by warmth, water, sun or wind".

Young people do manage to describe concepts of the formation or production of energy because this subject is omnipresent in the media with discussions on solar energy or wind energy being constantly mentioned on TV or radio broadcasts. In the students' minds, solar cells or windmills create energy – the fact that it comes from the sun's radiation or from mechanical energy through the movement of air is rarely mentioned in the media reports.

Students are even quite aware of the historical dimension of **thermal matter** – a 10th grade female student writes: "energy is an invisible matter with which one can, for instance, move (kinetic energy)", another pupil formulated: "energy is a substance which makes electricity appear". However, mass as a characteristic of a portion of matter or its density, is not transferred to energy portions. Finally, references are made to the common school experiment, i.e. the burning of steel wool on a triple-beam balance and they contain statements similar to the phlogiston theory (see Chap. 3): "steel wool becomes lighter, because weighable energy escapes".

Teaching and Learning Suggestions. Sketches of how to sensibly reflect on the concept of "thermal matter" with the help of experimental experiences are made; on the other hand, misconceptions regarding the "production and destruction of energy" should be abolished, whereas concepts on energy conversion should be constructed. Eventually, even the question of behavior, on the particle level, should be drawn to highlight possible explanations.

If forms of energy were to correspond with characteristics of matter, i.e. mass, then a closed apparatus in which energy is released to the surroundings after an exothermic reaction should become lighter. It is relatively simple for students to propose their own experiments related to this hypothesis. For instance, they could propose carrying out an exothermic reaction in a closed test tube or flask, whereby the mass would be measured before and after and then compared. An example would be to place several matches in a test tube closed off with a balloon and to weigh this assembly (see E10.1). After the match heads have been lit through heating, they could be weighed once again; no difference would be observed. The thermal and light energy released after the reaction of the matches causes no significant difference in mass. Having looked at similar experiments, it makes no sense to associate thermal energy with mass or to look at energy as "thermal matter".

A further experimental possibility would be the exothermic reaction of copper sulfate with water (see E10.2): the white copper sulfate is also closed off in the test

10.2 Empirical Research

tube with the balloon, which in this case also contains several drops of water. The white salt reacts spontaneously and visibly with the water to the blue hydrated copper sulfate. The advantage of this reaction is that no activation of energy is necessary. The chemical energy is transformed into thermal energy; the released energy can be directly felt with the hand or it can be measured with a thermometer as energy is transferred to the "cooling water" in a beaker (see E10.2).

The comparison of the mass of the closed test tube before and after and of the observation that the masses of the matter are identical before and after the reaction, shows that released energy portions do not add any mass – they do not represent any substance with the characteristic of "mass" and there is no "thermal matter".

There are two methods of indicating energy in reaction equations:

copper sulfate(s, white) + water(l) → copper sulfate hydrate(s, blue) + heat

copper sulfate(s, white) + water(l) →

copper sulfate hydrate(s, blue); heat is released

The first equation shows the amount of thermal heat in the sequence of reacting substances and could tempt the students to associate a type of "heat matter" in the reaction. The second equation clearly separates the substances from the energy by using a semicolon: the involved energy is indicated qualitatively through the terms "exothermic" or "$\Delta H < 0$", or they are quantitatively depicted in kJ per kg or in kJ per mol. The mixing of matter and energy in reaction symbols does not take place this way – and thus neither in the minds of the students.

In the last mentioned reaction of copper sulfate and water (see E10.2), the idea of a **conversion of one form of energy to another** could be looked at as the transfer of thermal energy to the cooling water in the beaker. This transferred energy portion can even be quantitatively ascertained: if the temperature of 20 g of cool water rises by 5°C, then we are dealing with an energy transfer of $E = 5 \times 20 \, cal = 100 \, cal$, i.e. energy of 418 J. The molar reaction enthalpy or the enthalpy per kg substance can be determined from these values, if the amount of white copper sulfate has been weighed.

Introducing chemical energy, which is concealed in white copper sulfate and which is released through the reaction with water to form blue copper sulfate, is much more difficult to teach. An energy diagram related to this phenomenon may help (see Fig. 10.3). If a certain thermal energy is added to a sample of blue hydrated copper sulfate (see E10.3), a system of white copper sulfate and water which contains higher chemical energy is formed. If white copper sulfate and water react (see E10.2), the difference in chemical energy is released through thermal energy and transferred to the cool water of the calorimeter.

In addition to thermodynamic considerations, discussions on the particles and their bonding could be added in more advanced lessons. The particles of hydrated copper sulfate absorb thermal energy and retain a higher kinetic energy: they separate from each other. The Differential Thermal Analysis

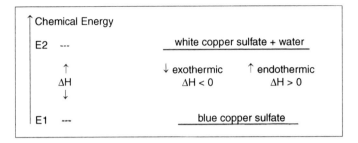

Fig. 10.3 Schematic energy diagram of dehydration of blue copper sulfate [11]

(DTA) is a method of analyzing such procedures: H_2O molecules of tetraaquacopper(II) complexes and those in the crystal lattice are not simultaneously separated from each other, but rather successively (see Fig. 10.4).

If one follows a DTA analysis with calcium sulfate dihydrate and of calcium sulfate semihydrate (gypsum), one gets the same results: H_2O molecules escape only through addition of a certain amount of thermal energy from the crystal lattice (see Fig. 10.5). Compared to an inert substance (like aluminum oxide) which heats up with a constant and almost linearly heating rate, the temperature recorder gives temperatures for two reactions while heating the calcium sulfate dihydrate: (a) at 140°C, 1 mol of the substance decomposes to 1 mol $CaSO_4$ ½H_2O and 1.5 mol H_2O molecules, (b) at 200°C, the last water is released. If one mixes the calcium sulfate semihydrate with water, the blend binds to solid crystals and is warmed at the same time: gypsum, as is well-known, becomes solid rather quickly upon addition of water through an exothermic reaction.

In any case, the idea of chemical energy should be intensified by showing many examples from everyday life; they should be linked to the **transformation of energy forms**. For instance, let us look at the example of the burning light bulb: the nearest power station transforms *chemical* energy of carbon through its strong exothermic combustion to *thermal energy*; this energy heats up steam and the *kinetic energy*, when released through the nozzles, drives the turbines and changes to *mechanical energy*. The turbine generator transforms the mechanical energy into *electrical energy*. At home, electrical energy can be transformed through the light

$$CuSO_4 \cdot 5H_2O \text{ (s)} \xrightarrow{93\ °C} CuSO_4 \cdot 3H_2O \text{ (s)} + 2H_2O$$

$$2H_2O \text{ (l)} \xrightarrow{102\ °C} 2H_2O \text{ (g)}$$

$$CuSO_4 \cdot 3H_2O \text{ (s)} \xrightarrow{115\ °C} CuSO_4 \cdot H_2O \text{ (s)} + 2H_2O \text{ (g)}$$

$$CuSO_4 \cdot H_2O \text{ (s)} \xrightarrow{250\ °C} CuSO_4 \text{ (s)} + H_2O \text{ (g)}$$

Fig. 10.4 DTA – diagram of dehydration of copper sulfate pentahydrate

10.3 Energy and Temperature

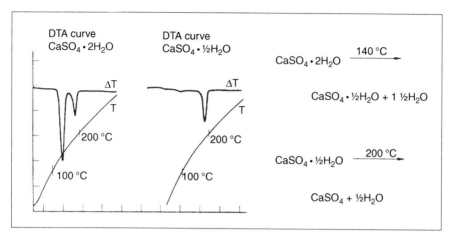

Fig. 10.5 DTA curves on decomposition of calcium sulfate dihydrate and semihydrate [11]

bulb into light energy and (unfortunately to a large extent) mainly into thermal energy. Other everyday examples that the students are exposed to are limitless.

It is open to discussion as to whether all these context-related examples will lead to the use of terms like power *transformation* instead of saying power *consumption*, if an exchange occurs from *fuel consumption* to *energy transformation* while speaking about driving a car. Changing to new descriptions is always a huge problem: we still hang on to old familiar expressions.

10.3 Energy and Temperature

The second and third questions to the students deal with energy and heating processes of water [10]. On the one hand, it's the change in the condition (the melting temperature) of ice which normally lies at 0°. On the other hand, students are asked what happens to water in a saucepan that completely evaporates after boiling for a while.

At first, an **ice–water mixture** at 0°C is introduced. The students are told that the mixture is heated for one minute with a burner, with both ice and water remaining. As possible answers, temperatures of the remaining ice–water mixtures are given: 0°C, 1°C, 5°C and −5°C. The first answer supplied in the multiple-choice question is correct. One student from the 10th grade gave the following adequate solution: "Energy is required to break up the crystal structure during the melting of ice; the temperature does not increase as long as the ice is not melted".

The statistical result of multiple-choice answers could be considered, good because 50% correct answers were given. However, even half of the students gave the correct explanation. Over 30% of all students chose the temperature increase of 1°C and explained: "The mixture must have become warmer because it was heated. It is, however, not much warmer because an ice–water mixture remains;

0°C must be incorrect because the water has been warmed up; the water is warmed up causing the ice to melt; because the water is warmed it becomes warmer; the fact that the ice is added means that it stays at 1°C". These or similar statements are given as reasons for the probable temperature increase of the ice–water mixture.

The learning theory of "Conceptual Growth" becomes quite clear at this point: students have obviously observed that temperatures of things increase when thermal energy is added – the existing cognitive structure supplies the basis of the mentioned preconcepts.

In the second exercise, the students have been given a description of the well-known observation, that a saucepan containing a certain amount of water becomes empty after being heated for a longer period. The following answers have been provided by the questionnaire: "water is burned; it reacts with the air; it decomposes to hydrogen and oxygen; water becomes gaseous and forms steam". The last alternative is of course the correct solution. A model drawing of the particle model of matter is also expected, depicting a mental models of the transition from the liquid state to the gaseous state.

On first glance, there is a satisfactory result in that the pupils mostly chose the correct answer, having chosen the formation of steam and given the explanation that water has a boiling point of 100°C – only 3% of students expect water to be broken down into its elements.

Usual model drawings are supplied which show the transition of the aggregation state from liquid to gaseous. However, reading the explanations show that the particle model and related model concepts on the aggregation states were mostly not fully understood: students mention only increasing distances between the particles, but not the increasing movement of particles. Some students also try to use the argument of repelling forces in between molecules instead of particle movement: "The attraction of the particles becomes increasingly diminished due to the heating process, until they finally repel each other and fly in the air; when the water is still cold, the water particles lay close together and at 100°C the particles fly away".

A few pupils discuss the result at the level of Dalton's atomic model by mentioning that steam is a mixture of hydrogen and oxygen. They state: "the steam is a sign of the reaction of H_2O to form H_2 and O_2: hydrogen and oxygen are gases, they are light and can disappear easily". It is also evident in the model drawings that pupils ponder on the splitting of both elements in the boiling of water process and postulate the occurrence of hydrogen and oxygen (see Fig. 10.6).

Finally, it is clear, that pupils cannot conceptualize the continual increase of the movement of particles, but rather imagine the boiling temperature as a point, which causes the particles to spontaneously fly away from each other "at the press of a button". Alternatively, they discover the model of the "expansion of the particles during the warming period", a pupil in the 10th grade writes: "I would say that the particles in water are arranged in a certain pattern and that they expand and burst when heated". It appears that there is a common connection in pupils' minds that "substances expand during the heating process":

10.3 Energy and Temperature

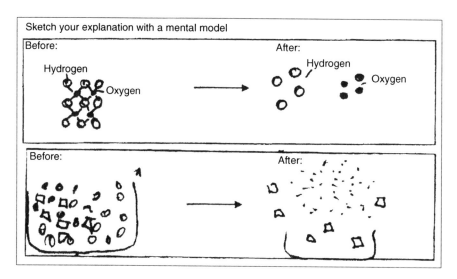

Fig. 10.6 Mental models concerning the boiling process of water [10]

they transfer these macroscopic properties subconsciously and unintentionally onto the sub-microscopic level of particles. With this, they are in good company as scientists of the past did the same [12].

Teaching and Learning Suggestions. The fact that ice melts at 0°C is well-known. Students are not so familiar with the fact that melting water has a temperature of 0°C, that for melting of 1 g of ice a very specific thermal energy has to be transferred, the **specific melting heat** of 333 J/g:

$$\text{ice (s, } 0°\text{C)} \rightarrow \text{water (l, } 0°\text{C)}; \Delta H = 333 \text{ J/g}$$

Led by the cognitive conflict that addition of energy always causes an increase in temperature, young people should certainly carry out their own experiments. On the one hand, they gradually add thermal energy to the ice–water mixture, while stirring, and they observe a constant temperature of 0°C (see E10.4). On the other hand, they recognize that it is irrelevant whether a little ice and a lot of water, or a lot of ice and a little water is in the beaker; a temperature of 0°C is always observed (see E10.4). If the idea for chemical equilibrium is supposed to be introduced with this example, it can be discussed through the addition or withdrawal of heat: each case will cause melting or solidification (freezing), respectively.

Through further heating of water to higher temperatures and the related transformation of chemical energy of fuel to thermal energy, students can learn about the **specific heat of water**. It is 4.18 J/gK and is interpreted as follows: the amount of energy of 4.18 J increases the temperature of 1 g water by 1°C (or by 1 K). One can show that a different liquid like, for instance, ethylene glycol with the value of 2.35 J/gK has a different specific heat: the smaller amount of energy of 2.35 J increases the temperature of 1 g glycol by 1°C (or by 1 K). Thus, the same amount of energy increases the temperature of a specific glycol portion to approximately double the temperature as a water portion of the same mass (see E10.5).

If one weighs a butane burner ("liquid gas") before heating up a portion of, for instance, 100 g water, and then heats up the water for a minute on wire gauze on a tripod, a specific maximum temperature increase can be determined (see E10.6). If one weighs the burner once again and uses a heat value for butane at 46000 kJ/kg, then one can determine which portion of the released chemical energy has been transferred to water. The missing part of the thermal energy – usually more than 80% – heats up the beaker, the wire gauze or the surrounding air (see E10.6). With this, the term of the efficiency level can be introduced and discussed and methods can be discussed to improve the efficiency level.

If one connects the particle model of matter with phenomena of the melting of ice or the boiling of water, it becomes more apparent to the students: in order to separate water particles from particle formation, special amounts of energy have to be spent. The particle movement increases with increased energy, specifically, energy-rich particles gradually release themselves from the particle formation – even below the boiling point. In this regard, one speaks of the evaporation of the water and of steam of water below boiling point.

In everyday language, the term "steam" is unfortunately also used to signify "haze", "fog" or "mist" – it is necessary to point out that "mist" is more a mixture of air and small water drops. Students may also think that water evaporates to form air – in the past this was quite common, steam was often identified with air [12]. In this sense it is important to note that gaseous steam condenses to liquid water (see also E10.2).

While simmering (just below boiling) at a constant normal pressure, the volume of a portion of matter increases drastically – examples such as simmering ethanol in a closed syringe or in a closed balloon can easily be demonstrated (see E4.9). With the mental model of increased free distances between the particles, volume increase is explained with the model experiment of pouring out closely packed spheres from a small glass bowl in a larger bowl followed by intensive shaking of these spheres (see E4.10). This model can clarify that the particles and their movements fill up the space, that particles that are moving intensely need a greater space than non-moving particles. In this sense misconceptions relating to an increase of the particle size with temperature should be discussed in detail and then completely changed into the scientific idea of moving particles. The fact that, unfortunately, air is found between the spheres in the model and that it is considered an "irrelevant ingredient", also makes it necessary to discuss this aspect: there is no air between the real particles; there is nothing (see Chap. 4).

10.4 Fuel and Chemical Energy

In three previous exercises, the subjects of burning steel wool and of burning carbon and gasoline have been dealt with. The students were supposed to choose correct answers that deal with the combustion reactions of exothermic reactions in which either energy is released or a part of the chemical energy is transformed to thermal energy. They were also expected to provide a model depiction along with the correctly chosen answer.

The **steel wool test** was used once again in order to clarify the possible concepts of destruction of matter (see Sect. 3.4) and to demonstrate energy transformation during combustion. Approximately half of the test students were able to correctly answer that "steel wool is heavier" after combustion than it was before. However, only a few of those questioned were able to provide an adequate reason or even draw appropriate model sketches.

Answers and explanations lead, on the one hand, to the fact that the masses are assumed to be identical before and after the burning process: "according to the law of conservation of mass, the mass of original matter is identical to the mass of final matter". Some answers show that "iron and oxygen are mixed in iron oxide". On the other hand, one is exposed to the well-known concept of "becoming lighter" which stems from the concept of destruction of matter: "during the burning of steel wool, certain elements are released from the wool; when something burns, it becomes lighter; particles are burned. Energy is released through combustion, this is why steel wool is lighter".

In one case, a complete diametrically opposite interpretation to the well-known oxidation theory is given: oxygen atoms are misplaced into the lattice of particles before combustion and released through the combustion process (see Fig. 10.7). In other cases, the usual "becoming lighter" phenomenon is associated with the release of energy: "energy is released through combustion and that's why the steel wool is lighter; when something burns, something is released and the weight is thereby diminished". Completely in line with this last exercise, a pupil from the 9th grade sketches his mental model (see Fig. 10.8): explanation and drawing point to the material characteristic of energy – to that of an imagined "heat matter" in Stahl's Phlogiston Theory of 1697 (see Chap. 1).

An above-average number of pupils have solved the exercise on the **combustion of carbon**; they also correctly described the reaction of carbon with oxygen to form carbon dioxide and determined that the combustion process supplies the energy. However, about 20% of students chose the distracter: "carbon is transmuted to energy, the result is ash".

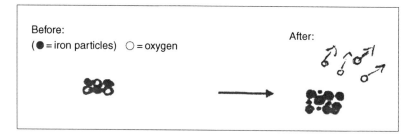

Fig. 10.7 Student's model drawing on the burning reaction of steel wool [10]

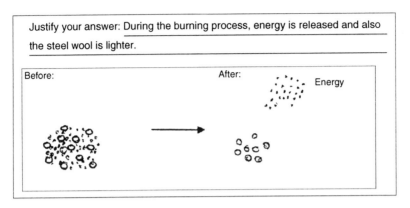

Fig. 10.8 Student's mental model on energy transfer during combustion of iron [10]

The result of the model drawing is more significant: less that 25% of the students were able to adequately supply a model concept. This is made evident by the performance of one 9th grade student: "C atoms are converted to energy and thus provide heat" (see Fig. 10.9). It is worth mentioning that students in grades 10 and 11 also think of this transmutation process from mass to energy as plausible.

A further obvious misconception related to this appears to be that superimpositions occur regarding the term activating energy. Students give various reasons for this. One student writes, "when carbon glows, it stores energy and warmth, but after some time it turns to ashes and releases energy once again". Another student formulates, "when carbon is lit, energy is added which is later released".

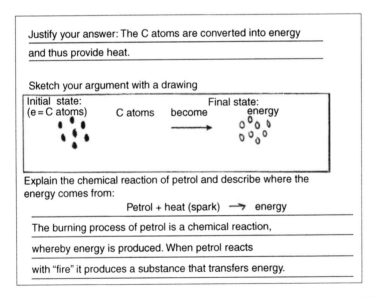

Fig. 10.9 Students' mental models on combustion of carbon and gasoline [10]

10.4 Fuel and Chemical Energy

One type of fuel, which most students are familiar with, is petrol, which explodes in the mixture with oxygen and causes the motor to run. The combustion processes are not immediately as visible as those of the visible combustion of charcoal are – this is why it is interesting to create an exercise using this example. Many students supplied acceptable solutions concerning the reaction of gasoline (petrol) with oxygen, stating that it is an exothermic reaction. Only a few found the correct equation, mentioning the formation of carbon dioxide and water. Many students only use the carbon dioxide argument: "petrol is a carbon compound, the C atoms react to CO_2, the other atoms transform into either a gaseous or a solid matter".

However, in the explanation, very few statements correctly approach the matter from the combustion reaction aspect that chemical energy transmutes to mechanical energy – only very few pupils are aware of this. In most cases, the usual generation of energy from petrol is at the forefront of their ideas: "energy is created during combustion, the petrol atoms can react with oxygen during the combustion process, thereby forming energy; during a car trip, petrol is burned, it diminishes and after about 200 km one has to fill up the tank again; that which comes out of the exhaust is gaseous, burned petrol" [10]. One grade 10 student even creates the equation, "fuel + heat → energy", depicting this process as if it is obvious, between the terms "matter" and energy, and between "heat matter" and energy (see Fig. 10.9).

The majority of students stress that explosions take place in the car's engine. This is true, however, it appears as if the real reasons for such explosions have not been understood – most students do not think of reactions of a petrol-oxygen mixture. It appears more as if petrol just spontaneously explodes and such explosions automatically lead to the well-known misconception of matter destruction: "petrol disappears during the explosion and it is necessary to fill up petrol again at some stage" [10].

Teaching and Learning Suggestions. The combustion of iron and mental models related to this were discussed in Chap. 3. The explanation regarding the burning of charcoal is difficult: no solid oxide is released as in the case of iron, but rather a gaseous and invisible oxidation product. In addition, the burning process is accompanied by remnants of white ash and leads the observer to assume this as being the possible reaction product. In addition, petrol does not visibly burn in air to carbon dioxide and steam. However, the visible result is black soot that, like the ash produced by burning charcoal, could be interpreted as a reaction product. Therefore, both cases are rather difficult to deal with from a chemistry didactic point of view: the actual products being dealt with are invisible, the accompanying products are visible but that is not what interests us.

About the burning of carbon, it should be made clear that pure carbon, when reacting with oxygen, leads completely to a colorless carbon dioxide gas without leaving any remnants. In an apparatus with two syringes and a combustion pipe, several pieces of carbon are placed into the combustion pipe and one syringe is filled with oxygen (see E10.7). After the intense heating of the pipe, the carbon pieces, after oxygen has been added, react completely producing a bright light – without leaving any remnants. The gas volume remains constant because the oxygen is replaced by the same amount of carbon dioxide. The existence of this gas can be observed using limewater. The reaction should be demonstrated at first with the use of a molecule building set and afterwards be combined with the well-known reaction equation. Even the constant volume of the involved gases can be interpreted in this connection, particularly if Avogadro's Law is supposed to be introduced.

The phenomenon of leftover ashes in the burning of charcoal can be explained through a kind of contamination: minerals like white sodium carbonate, white potassium carbonate ("potash") and sulfur are, among others, contained in common coal. If connections are made to the historic development of brown coal or hard coal from the remains of plants, it appears to be quite clear, that coal extracted in mining cannot be a pure product.

Regarding the question of energy release, it is necessary for the students to raise the question of chemical energy and to transfer it to the well-known diagram (see Fig. 10.3): the carbon–oxygen mixture is, in comparison to carbon dioxide, a substance system containing relatively high chemical energy. During the reaction, the correlative energy difference is released to the surroundings: chemical energy of 1 kg of carbon leads to a thermal energy of 33,000 kJ during a complete reaction with oxygen in the air. If one looks at the reaction closely,

$$C(s) + O_2(g) \rightarrow CO_2(g); \Delta - H = -33000 \, kJ/kg$$

it is possible to qualitatively distinguish the reaction equation in the following way:

1. C atoms have to be separated from the atomic lattice by the expenditure of energy,
2. O_2 molecules should be separated to O atoms by the use of energy,
3. C atoms and O atoms combine to CO_2 molecules by gaining energy.

However, before the first individual C atoms or O atoms are separated, it is not possible to start the reaction. For this reason, it is necessary to clarify using first-hand experiences, that a piece of coal does not start burning on its own, but rather a certain "initial energy" or "activation energy" is required in order to continue the reaction. After this initial start, any amount of coal can be added and will burn as long as coal is present.

Quantitatively, such reactions can be described by using the bond energy of the C-C bond, the C-H bond, the C-O bond and O-H bond. Knowing these values, appropriate reaction enthalpies can be calculated. A clear example of the formal reaction of graphite, hydrogen and oxygen to ethanol is described by Dickerson–Geis (see Fig. 10.10).

In addition, **petrol**, as a mixture of different hydrocarbons, is a substance of relatively high chemical energy. As a carbon and hydrogen compound, it reacts with oxygen to form carbon dioxide and steam, two substances of relatively low

10.4 Fuel and Chemical Energy

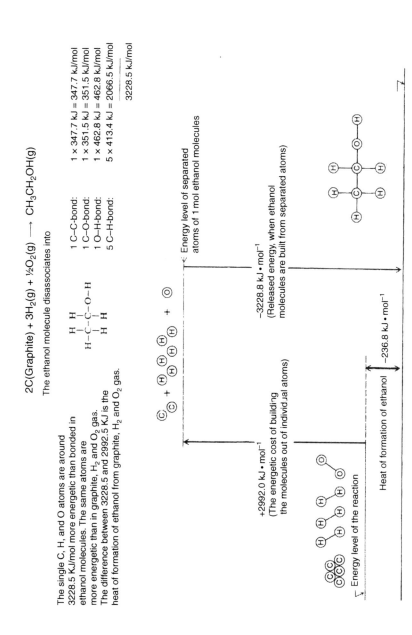

Fig. 10.10 Model depiction on energy turnover in the synthesis of ethanol [13]

energy (see Fig. 10.3). These reaction products can be proven – also quantitatively – using by the butane–oxygen reaction in an experiment (see E10.8). In order to obtain further quantitative indications regarding the appearance of colorless and gaseous reaction products, the combustion products could be combined with absorbent substances and weighed. One possibility for such an experiment is to burn a paraffin candle – another mixture of hydrocarbons like in petrol – on the weighing balances (see E3.9).

The heat and light energy released in the petrol reaction is made evident through the bright yellow flame that most students are familiar with. The formation of soot through the incomplete burning in air is well-known: small crystals of graphite and other C atom conglomerates are released in the air and form a not-so-pleasant fine dust in cities with heavy traffic.

Not so well-known – because it is invisible – is the explosive energy conversion of petrol–oxygen mixtures, which can be modeled in the processes that occur in a car engine. One safe possibility of demonstrating the workings of a car engine is to fill a plastic bottle with the above-mentioned mixture and to light it with a car spark plug (see E10.9). The small plastic bottle, whose opening fits on the base of the spark plug, is filled with a few drops of pentane or petrol and with oxygen. After shaking in a warm hand, it is placed on the spark plug and ignited. The loud bang – and the shredding of the plastic bottle – is due to the huge amount of chemical energy that is stored in the mixture of a mere few drops of petrol and oxygen. By igniting the mixture, the chemical energy is transformed into both thermal energy and light energy.

In order to make this process more understandable, the molecular building set can be used (see E10.10): the structure model of a pentane molecule, C_5H_{12}, should be constructed and also the isomers, the reaction of pentane with oxygen can be modeled practically or mentally by the use of O_2 molecules models. The students should learn that a maximum of 5 CO_2 molecules and 6 H_2O molecules can result from one C_5H_{12} molecule. The reaction equation can only be properly understood after carrying out this reflection on the amount of related molecules:

$$1\ C_5H_{12}\ \text{molecule} + 8\ O_2\ \text{molecules} \rightarrow 5\ CO_2\ \text{molecules} + 6\ H_2O\ \text{molecules; exothermic}$$

Regarding engine function, the conversion of chemical energy of petrol can be discussed. Through the reaction of petrol and oxygen, the released chemical energy is transformed into thermal energy causing the large volume expansion of the gaseous reaction products, carbon dioxide and steam. The resulting mechanical energy causes the pistons of the engine to move up and down through the special construction of the motor. The mechanical energy produced by the rapid up- and downward motion of the pistons causes the car wheels to turn or perhaps drives an alternator via transmission belts. These again transform mechanical energy into electrical energy and move the starter or the light bulb: in this way, light energy is another conversion form of chemical energy of petrol and oxygen. However, through the increasing advancement in automobile technology, many more energy forms play a role.

Finally, one can point out that not all thermal energy can be transformed completely into mechanical energy or light energy: certain portions of this energy lead to an increase in entropy. If one wishes to show chemical reactions with complete conversion of chemical energy in electrical energy, one has to perform experiments in electrochemistry. On the one hand, the galvanic cell itself (see E8.7) shows this conversion; on the other hand, a model experiment on the lead accumulator (see E8.10) could demonstrate by loading and unloading energy in both directions. These and further discussions with students will develop a better "scientific literacy" on energy and the conversion of different forms of energy.

10.5 Experiments on Energy

E10.1 Masses and Exothermic Reactions

Problem: Pupils may have the opinion that there is such a thing as "heat matter", or that different samples of energy exhibit different masses. In an exothermic reaction, the involved substances should have to lose a portion of their mass – the law on conservation of mass in chemical reactions would not apply in this case. In order to remind students of this law, an exothermic reaction should be quantitatively introduced.

Material: Test tubes, balloon, balances; matches.

Procedure: Place several match heads in a test tube, close it with a balloon and weigh it. Heat the test tube with the hottest flame of the burner until match heads are lit. After cooling down weigh the test tube and content once more.

Observation: The masses are identical; the energy portion that is released as thermal energy and light energy has no mass.

Tip: Ever since the conception of the Relativity Theory of Einstein's formula, $E = mc^2$, one assumes, that the *sum* of mass and energy remain constant during chemical reactions. Thereby, mass, which is equivalent to the energy demonstrated in chemical laboratories, is so small that it cannot be captured by any scales. Only through the release of enormous energy in nuclear fission or fusion reactions is it possible to obtain weighable amounts of energy. For all practical purposes, weighing these is not possible either.

E10.2 Energy Transfer During Exothermic Reaction

Problem: Apart from misconceptions on "heat matter", it is the "origin of energy" or "consumption of energy" which pupils' encounter in everyday language. Therefore, this should be shown through weighing the conservation of mass in a reaction. Additionally, the transformation of chemical energy to thermal energy should be demonstrated in a calorimeter. After all, the transferred thermal energy can be computed through the known mass of the cooling water and the measured difference in temperatures.

Material: Styrofoam cup (calorimeter), test tubes, balloons, measuring cylinder, balances, thermometer; white copper sulfate.

Procedure: Measure exactly 2.0 g of white copper sulfate in a test tube, close the tube using a balloon that contains about 2 ml water. Fill exactly 20 ml of cooling water into the styrofoam cup and measure the temperature of the water. Place the test tube in the calorimeter, transfer the water held in the balloon into the test tube by lifting the balloon and measure the temperature of the cooling water until a constant temperature is attained.

Observation: The white salt changes to a blue-colored salt, the temperature of the cooling water rises by about 5°C.

Tip: The increase of water temperature by 1°C means an addition of thermal energy of 4.18 J for 1 g water. If 20 g of water are used and the temperature difference of 5°C is observed, then an energy amount of $4.18 \, J \times 20 \times 5 = 418 \, J$ has been transferred. The reaction enthalpy can be calculated by 1 kg or 1 mol of copper sulfate.

E10.3 Energy Transfer During Endothermic Reaction

Problem: The blue substance formed from E10.2 originates from the white copper sulfate and water; it contains water of hydration (or crystallization) and is called copper sulfate pentahydrate. Apparently, this blue copper sulfate has a lesser amount of chemical energy than the white copper sulfate (see Fig. 10.3). Adding thermal energy, the blue sulfate can be converted into white sulfate with higher chemical energy.

Material: Test tubes, burner; blue copper sulfate (see E10.2).

Procedure: Fix the test tube containing the blue copper sulfate in a slanted position with the opening pointing downward and heat with the hottest burner flame.

Observation: The blue sulfate changes to white sulfate, drops of water condense and drip out of the test tube. As soon as the heat supply is interrupted, the decomposition stops.

E10.4 Temperatures of Ice–Water Mixtures

Problem: Young people know from experience, that when heating water, each addition of thermal energy leads to an increase in temperature – temperature and heat have the same meaning for many pupils. For this reason, experiments should be done and observations and results evaluated which show constant temperatures during melting of pure materials: for instance, the simple melting of ice at 0°C to water with 0°C.

Material: Beaker, tripod and wire gauze, thermometer; ice.

Procedure: Measure the temperature of ice that has been freshly removed from the refrigerator. Heat the ice using a burner while constantly stirring using a thermometer. Measure the temperature regularly at certain intervals. Intermittently, add different portions of ice or water to this mixture, stir and measure the temperature when it stays constant.

Observation: The ice that has just been taken from the ice compartment has temperatures under 0°C. All other ice–water mixtures show a temperature of 0°C.

Tip: The fact that all ice and water mixtures – independently of the mass of ice and water – show a temperature of 0°C, can lead to the concept of chemical, dynamic equilibrium of forward and backward reaction (see Chap. 6).

E10.5 Specific Heat of Water and Glycol

Problem: Concepts of energy and temperature have not been adequately differentiated in students' minds – heating with identical flame for identical periods of time leads, in their minds, to an identical temperature increase, if same portions of different substances have been chosen. But for exact same portions of two different liquids with clear differences in specific heat (4.18 J/gK for water, 2.35 J/gK for glycol), identical amounts of energy leads to two completely different temperatures. If one takes a butane burner with a constant flame and constant period of time, the burner delivers the same amount of energy. This amount of energy is transferred to portions of water and glycol of the same masses, and the temperatures are compared.

Material: Two identical 250 ml beakers, butane burner, tripod and wire gauze, thermometer, balances, stopwatch; water, ethylene glycol.

Procedure: Take two beakers and fill one with 100 g of water and the other with 100 g of ethylene glycol. Measure temperatures of both liquids. Place wire gauze on the tripod and heat with a burner adjusting a constant heat supply. Place the first beaker for 60 s on the heated wire gauze, then the second one. Measure the maximum temperature reached in each beaker.

Observation: The temperature of the water increases to about 15°C while glycol temperature increases to about 28°C.

Tip: The differences of water and glycol in relation to their specific heat have an influence on the cooling properties for car engines: pure water absorbs larger amounts of energy than mixtures of water and glycol (Glysantine). However, these mixtures have to be used to avoid water freezing in car coolers in winter.

E10.6 Efficiency Level of Combustion

Problem: In all experiments during which liquid is heated using the tripod and the wire gauze, pupils do not consider the fact that a large portion of the supplied thermal energy is not effectively transferred onto the liquid, but rather leads to the heating up of whatever material happens to be used, i.e. glass (beaker), iron (tripod) and the surrounding atmosphere (air). This fact is associated with the term "efficiency", which could be added for discussion.

Material: 250 ml beaker, butane burner, tripod and wire gauze, thermometer, balances, stopwatch; water.

Procedure: Put 100 g of water in a beaker, place the beaker on the wire gauze on the tripod and measure the temperature. Exactly weigh the mass of the butane burner, turn it on and adjust the hottest burner flame, immediately position it under the beaker and start the stopwatch. After exactly 60 s, remove the burner, turn it off immediately and weigh the burner again. Record the maximum temperature attained by the water.

Observation: By the reaction of 1.0 g of butane the water temperature rises by 15°C.

Tip: The efficiency level is estimated as follows: tables show the heat value of 46000 kJ per 1 kg of "liquid gas"; this would mean that 1 g of that fuel supplies the thermal energy of 46 kJ. 100 g of water use this amount of energy: $100\,g \times 15\,K \times 4.2\,J/gK = 6270\,J = 6.27\,kJ$ for a temperature increase of 15°C. This amount of energy correlates to 13.6% of 46 kJ – therefore the efficiency level is only 13.6%: more than 86% of the produced thermal energy is released into the environment (glass, iron, air, etc.).

E10.7 Quantitative Combustion of Pure Carbon

Problem: After a barbeque, there is always a remnant of white ash – it is a bit of a nuisance having to dispose of it. Young people are often called to assist in this task. They tend to believe, on the one hand, that ash is the product of burning coal, and that ash is "lighter" afterwards than coal was before. On the other hand, many students believe in a kind of "heat matter", which is released whenever combustion takes place. This misconception could be corrected with this experiment.

Using pure carbon may show that, on the one hand, no ash is formed and, on the other hand, that in the exothermic reaction with oxygen is delivered gaseous, colorless carbon dioxide along with high amount of thermal energy: 31000 kJ per kg carbon. By the way, carbon dioxide as the combustion product takes the same volume as oxygen before the reaction (at normal pressure): every O_2 molecule is replaced by one CO_2 molecule, the number of molecules does not change the volume stays constant at the same pressure and same temperature (Avogadro's law).

Material: Two 100 ml syringes, combustion tube, small beaker; chunks of activated carbon, calcium hydroxide solution (limewater), oxygen (lecture gas bottle).

Procedure: Horizontally position the combustion tube containing some pieces of activated carbon. Connect it to two syringes; one is filled with 100 ml of oxygen, and the other syringe is empty. Heat the carbon pieces until glowing with the hottest burner flame, and then pass oxygen slowly over it. At the end of the reaction, bubble a portion of the contents of the produced gas into a few mls of limewater.

Observation: The carbon pieces light up and show a strong exothermic reaction with oxygen, until they completely disappear. The volume of 100 ml gas in the piston sampler remains intact. When the gas reacts with the limewater, a white deposit precipitates.

Tip: In order to show the conservation of mass, it is possible to take a big flask, fill it with oxygen, add some kernels of activated carbon, close it with a stopper and weigh it exactly. With the burner, the carbon is heated until it glows; the flask is taken away from the burner. When the carbon pieces "disappear", the flask is weighed one more time: the same mass will be observed. With the help of limewater, the presence of carbon dioxide may be confirmed.

Inexpensive industrial diamonds, available from educational suppliers, could also be used in order to show that diamonds are chemically identical with carbon, that they are composed of C atoms and react with O_2 molecules to form CO_2 molecules. The observations are identical to those described above.

E10.8 Quantitative Oxidation of Butane

Problem: As in the case of carbon combustion, the gaseous products from the combustion of butane, pentane or gasoline are not noticeable – the opposite is in fact the case: gasoline burning in a porcelain bowl leaves a black sooty deposit and young people may interpret the soot as a combustion product. In order to circumvent the soot formation, gaseous butane is used in this experiment. It is possible to qualitatively prove the existence of steam and carbon dioxide when examining the butane burner flame. Quantitative testing for carbon dioxide with the help of copper oxide is also possible.

Material: Washing bottle, water aspirator (or vacuum pump), funnel, two syringes, combustion tube, small beaker, butane burner; black copper oxide (wire form), glass wool, limewater.

Procedure: (a) Adjust a small burner flame; attach a funnel over the flame and connect a wash bottle which has been filled with a few mls of limewater. Mount the water aspirator to the correct end of the wash bottle, and suck the combustion products through the bottle. (b) Put several spoonfuls of copper oxide into the combustion tube and hold it with pieces of glass wool. Fill one syringe with 20 ml of butane from the butane burner and attach to one side of the combustion tube; connect the other syringe to the opposite side of the tube opening. Heat the copper oxide with the hottest burner flame, and then slowly pass the butane over the glowing copper oxide several times until the gas volume remains constant. Bubble the gas through a small amount of limewater in a beaker.

Observation: (a) At the beginning, one can clearly see a water deposit in the wash bottle; the limewater delivers the usual white deposit: steam and carbon dioxide. (b) 80 ml of a colorless gas are formed; the gas produces a white deposit in limewater: carbon dioxide.

Tip: The increase in volume of 20 ml of butane to 80 ml of carbon dioxide is interpreted as follows: 1 mol butane molecules delivers 4 mol CO_2 molecules, so one molecule of butane has to contain 4 C atoms: the formula C_4H_{10} results. The produced steam condenses during the reaction to several drops of water, they cannot account for volume calculation.

The combustion products of all hydrocarbons can be identified as carbon dioxide and steam, the origin of the soot deposit could be discussed: their incomplete, too rapid combustion produces black carbon, described as soot, and is dispersed in the air. Even a candle, which initially burns without producing visible soot, produces a lot of black soot when a porcelain bowl is held over the flame, preventing it from burning properly.

E10.9 Explosion of Gasoline–Oxygen Mixtures

Problem: Most young people are aware that gasoline is burned in car engines and that the transferred energy causes the moving of pistons and thus the car to move. They are most probably also aware of the existence of spark plugs. Less known to these young people, however, is the fact that gasoline has to first evaporate and be mixed with air before an ignitable gas mixture can cause explosions. Both of the following should be demonstrated in this experiment: the mixing of oxygen with fuel vapor and the igniting of the gas mixture with a spark plug. Due to the loud bang, it is easy to demonstrate the large amount of chemical energy in the gasoline–oxygen system that has been transformed to thermal energy. It should be pointed out that the exothermic reaction of hydrocarbons with oxygen forms carbon dioxide and steam (see also E10.8).

Material: 100 ml plastic bottle (the opening has to fit on the base of a spark plug), spark plug, spark igniter (or Tesla coil), cable; gasoline (pentane), oxygen (lecture gas bottle).

Procedure: Position the spark plug vertically with its base pointing upwards; connect both poles to the spark igniter. First test if an ignition takes place. Fill the plastic bottle with oxygen, add about three drops of gasoline and close the bottle with the thumb. While warming the contents of the bottle with hands, the gasoline–air mixture is produced. Tightly screw the bottle opening to the base of the spark plug. Keeping a safe distance, press the igniter (Caution: loud bang, destroyed plastic bottle catapults vertically to the ceiling!).

Observation: A loud bang results, the bottle is propelled away from the spark plug, it lands, completely in shreds, on the floor.

Tip: A small plastic bottle should certainly be used, never a glass container or another solid material as this could easily cause injury. It is also possible to take a small cylinder, produce the gasoline–oxygen mixture and ignite it with a burning wooden splint. But again, don't do it with a glass round flask that may explode and produce glass shards.

E10.10 Molecular Models for Hydrocarbon–Oxygen Reactions

Problem: Given the names carbon dioxide and steam, which are not even visible after the reaction of hydrocarbons with oxygen, do not seem capable of convincing students, who have internalized the destruction concept. They may have studied the law of conservation of masses during chemical reactions – but they do not believe it in the case of burning gasoline.

A molecular building set is suitable for representing the molecule structures of pentane and oxygen molecules before the reaction. They, then, could build the molecule models for the reaction products, carbon dioxide and water, through rearranging the atoms from the initial models. Using the models, students could better understand, in this example, that a maximum of 5 CO_2 molecules and 6 H_2O molecules could be formed from one C_5H_{12} molecule, and that 8 O_2 molecules are needed for this. The quantitative reactions of carbon and butane with oxygen (see E10.7 and E10.8) are also better understood with the molecular building set. Avogadro's law should also be taken into consideration if gas volumes are to be quantitatively compared.

Material: Molecular building set.

Procedure: Construct the model of one C_5H_{12} molecule and position it on one side of the table together with models of O_2 molecules; on the other side of the table, place models of CO_2 and H_2O molecules. Ask how many molecule models are on both sides of the table: $5CO_2$ molecules can be formed from 5 C atoms in one pentane molecule; 6 H_2O molecules can be formed from 12 H atoms in one pentane molecule. Finally, a counts of 8 O_2 molecular models are needed.

Observation: By counting all models, one can state that 1 C_5H_{12} molecule will react with 8 O_2 molecules to form 5 CO_2 molecules and 6 H_2O molecules.

Tip: Based on the model reaction, the correlative reaction equation should be worked out on the blackboard, using structural symbols of the involved molecules first, and then shortening them into usual molecular formulas. The energy turnover could be represented through dismantling (breaking) and resetting up (forming) bonds between the atoms: the breaking of bonds uses energy, the formation of bonds supplies energy. Finally, models could also represent isomer molecules of pentane.

Because the majority of organic substances are made up of molecules, the molecular building set could further be used to demonstrate the many areas in organic chemistry. This is why organic chemistry is often very popular: young people gain a better comprehension by using the molecular building set. The structure-oriented approach should be part of every curriculum in chemistry, students would understand chemistry much better!

References

1. Scheler, K.: Energie als Tauschwert – ein neuer Ansatz zur Erschließung des Energiebegriffs in der Sekundarstufe I. Chim. did. 30 (2004), 67
2. Pfundt, H., Duit, R.: Bibliographie – Alltagsvorstellungen und naturwissenschaftlicher Unterricht. Kiel 1994 (IPN)
3. Burger, J., Gerhardt, A.: Energie im biologischen Kontext. MNU 56 (2003), 324
4. Barker, V.: Beyond Appearances: Students' Misconceptions About Basic Chemical Ideas. London 2000 (Royal Society of Chemistry)
5. Driver, R., Guesne, E., Tiberghien, A.: Children's Ideas in Science. Philadelphia 1985 (Open University Press)
6. Erickson, G., Tiberghien, A.: Heat and temperature. In: Driver, R., et al: Children's Ideas in Science. Philadelphia 1985 (Open University Press)
7. Kircher, E., Schneider, W.B.: Physikdidaktik in der Praxis. Heidelberg, New York 2003 (Springer)
8. Schmidt, H.J.: Problem Solving and Misconceptions in Chemistry and Physics. Dortmund 1994 (ICASE)
9. Griffiths, A.K.: A critical analysis and synthesis of research on students' chemistry misconceptions. In: Schmidt, H.J: Problem Solving and Misconceptions in Chemistry and Physics. Dortmund 1994 (ICASE)
10. Doerfler, T.: Brennstoffe und Energie – Empirische Erhebungen zu Schuelervorstellungen und Unterrichtsvorschlaege zu deren Korrektur. Staatsexamensarbeit. Muenster 2004
11. Wiederholt, E.: Differenzthermoanalyse (DTA) im Chemieunterricht. Koeln 1981 (Aulis)
12. Schleiß von Loewenfeld: Anfangsgründe der Physik. Muenchen 1861
13. Dickerson, R.E., Geis, I.: Chemie – eine lebendige und anschauliche Einfuehrung (Chemistry, Matter and the Universe). Weinheim 1981 (Verlag Chemie)

Further Reading

Beall, H.: Probing student misconceptions in thermodynamics with in-class writing. Journal of Chemical Education 71(1994), 1056

Boo, H.K.: Student understandings of chemical bonds and the energetics of chemical reactions. Journal of Research in Science Teaching 35 (1998), 569

Boohan, R., Ogborn, J.: Energy and Change Association for Science Education. Hatfield 1996

Cachapaz, A.F., Martins, I.P.: High school students' ideas about energy of chemical reactions. In: Novak, J., Helm, H.: Proceedings of the International Seminar on Misconceptions in Science and Mechanics 3 (1987), 60

Clough, E., Driver, R.: Secondary students' conceptions of the conduction of heat: Bringing together scientific and personal views. Physics Education 20 (1985), 176 de Vos, W., Verdonk, A.: A new road to reactions. Part 3: Teaching the heat effect of the reaction. Journal of Chemical Education 63 (1986), 972

Duit, R.: Learning the energy concept in school – empirical results from the Philippines and West Germany. Physics Education 19 (1984), 59

Finegold, M., Trumper, R.: Categorizing pupils' explanatory frameworks in energy as a means to the development of a teaching approach. Research in Science Teaching 19 (1989), 97

Johnstone, A.H., MacDonald, J.J., Webb, G.: Misconceptions in school thermodynamics. Physics Education 12 (1977), 248

Kesidou, S., Duit, R.: Students' conceptions of the Second Law of thermodynamics, an interpretive study. Journal of Research in Science Teaching 30 (1993), 85

Lewis, E., Linn, M.: Heat, energy and temperature. Concepts of adolescents, adults and experts: implications for curriculum development. Journal of Research in Science Teaching 31 (1994), 657

Ross, K.: There is no energy in food and fuels – but they do have fuel value. School Science Review 75 (1993), 39

Shipstone, D.M.: A study of children's understanding of electricity in simple DC circuits. European Journal of Science Education 6 (1984), 185

Solomon, J., Black, P., Oldham, V., Stuart, H.: The pupil's view of electricity. European Journal of Science Education 7 (1985), 281

Solomon, J.: Teaching the conservation of energy. Physics Education 20 (1985), 165

Trumper, R.: Being constructive: An alternative approach to the teaching of the energy concept. International Journal of Science Education 12 (1990), 343

Trumper, R.: Children's energy concepts: A cross-age study. International Journal of Science Education 15 (1993), 139

Trumper, R.: A survey of the conceptions of energy in Israeli pre-service high school biology teachers. Journal of Science Education 19 (1997), 31

List of Experiments

3.7	**Experiments on Substances and Their Properties** 52
E3.1	Heating a Copper Envelope 52
E3.2	Heating of Copper in Vacuum and in Air 53
E3.3	Decomposition of Silver Sulfide or Silver Oxide 53
E3.4	Reaction of Copper Oxide with Hydrogen 53
E3.5	Evaporation of Acetone 54
E3.6	Condensation of Butane Gas Under Pressure 54
E3.7	Dissolution of Metals 55
E3.8	Dissolution of Grease 56
E3.9	Burning Metals on a Balance 57
E3.10	Conservation of Mass by Burning Charcoal 58
E3.11	Burning Candles on a Balance 58
E3.12	Reaction of Carbon Dioxide with Magnesium 60
E3.13	Density of Air and Carbon Dioxide 61
E3.14	Properties of Hydrogen and Other Colorless Gases 61
E3.15	Composition of Air 63

4.8	**Experiments on Particle Model of Matter** 93
E4.1	Growing of Alum Crystals 93
E4.2	Close-Packing Model for the Alum Crystal 94
E4.3	Electrostatic Forces for a Bonding Model 94
E4.4	Silver Crystals from a Silver Salt Solution 94
E4.5	Cubic Close Sphere Packings as Models for a Silver Crystal .. 95
E4.6	Solution of Iodine in Ethanol 96
E4.7	Sphere-Model for the Solution of Iodine in Ethanol 96
E4.8	Reaction of White and Red Phosphorus with Oxygen 97
E4.9	Volumes of Liquid and Gaseous Ethanol 97
E4.10	Model Experiments for the Three States of Matter 98
E4.11	An Empty Flask is Full of Air 98
E4.12	A Flask Contains Nothing 99

5.5 Experiments on Structure–Property Relationships 130
- E5.1 Silver Crystals Through Electrolysis 130
- E5.2 Closest Sphere Packings as Models for Metal Crystals 131
- E5.3 Cubic Closest Packing as a Model for Silver Crystals 132
- E5.4 Ductility of Different Metals 132
- E5.5 Formation and Decomposition of Sodium Amalgam 133
- E5.6 Nitinol – A Memory Metal 134
- E5.7 Freezing Point Temperatures of Solutions 135
- E5.8 Structural Models for the Sodium Chloride Structure 135
- E5.9 Destruction of Salt Crystals 137
- E5.10 Electric Conductivity of Salt Crystals, Melts and Solutions . 137
- E5.11 Heat of Crystallization from Molten Salt 138
- E5.12 Precipitation of Salt Crystals from Solutions 138
- E5.13 Electric Attraction and Repulsion Forces 139
- E5.14 Ionic Bonding and Magnetic Model 139
- E5.15 Magnetic Force Fields 140

6.4 Experiments on Chemical Equilibrium 165
- E6.1 Melting Equilibrium 165
- E6.2 Solubility Equilibrium of Salt 165
- E6.3 Model Experiment on Dynamic Equilibrium 166
- E6.4 Solubility Equilibrium of Sodium Chloride 167
- E6.5 Solubility Equilibrium of Calcium Sulfate 167
- E6.6 Boudouard Equilibrium 168
- E6.7 Quantitative Decomposition of Ammonia 168

7.4 Experiments on Acids and Bases 193
- E7.1 Decomposition of Sugar Using Sulfuric Acid 193
- E7.2 Reaction of Metals with Sulfuric Acid 193
- E7.3 Limestone Deposit Removers – Acidic Household Cleaners ... 194
- E7.4 Drain Cleaner – an Alkaline Household Chemical 194
- E7.5 Reaction of Several Acid–Base Indicators 195
- E7.6 pH Values of Several Bathroom and Kitchen Chemicals 196
- E7.7 Electrical Conductivity of Solutions of Acids, Bases and Salts .. 196
- E7.8 Hydrochloric Acid – from Solid Sodium Chloride 197
- E7.9 Hydrochloric Acid – from Gaseous Hydrogen Chloride 197
- E7.10 Sulfuric Acid and Water – a Strong Exothermic Reaction ... 198
- E7.11 Reaction of Calcium Oxide and Water 198
- E7.12 pH Values – by Dilution of Hydrochloric Acid 199
- E7.13 Reaction of Calcium Hydroxide Solution with Carbon Dioxide 199
- E7.14 Neutralization Reactions 200
- E7.15 Conductivity Titration of Baryta Water with Sulfuric Acid Solution 201
- E7.16 Heat of Neutralization 202

List of Experiments

| | E7.17 | pH Values of Strong and Weak Acids | 202 |
| | E7.18 | pH Values of HCl and H_2S Solutions | 203 |

8.3 Experiments on Redox Reactions ... 226
- E8.1 Precipitation of Copper from Copper Sulfate Solution ... 226
- E8.2 Exothermic Precipitation Reactions ... 227
- E8.3 Precipitation of Silver from Silver Nitrate Solution ... 227
- E8.4 Precipitations with Other Metals ... 228
- E8.5 Metals in Acidic Solutions ... 228
- E8.6 Metals in Salt Solutions ... 228
- E8.7 Galvanic Cells ... 229
- E8.8 Iron Corrosion ... 230
- E8.9 Leclanché Battery ... 230
- E8.10 Lead Accumulator ... 230

9.3 Experiments on Complex Reactions ... 252
- E9.1 Existence of Complex Ions ... 252
- E9.2 Blue and White Copper Sulfate# ... 252
- E9.3 Octahedral Models of Complex Particles ... 253
- E9.4 Properties of the Ammine Copper Complexes ... 254
- E9.5 Determination of the Coordination Number ... 254
- E9.6 Copper Complexes and Stability ... 255
- E9.7 Nickel Complexes and the Octahedral Structure ... 256
- E9.8 Synthesis of Hexa Ammine Cobalt Chloride Crystals ... 256
- E9.9 Copper Complexes and Equilibrium ... 257
- E9.10 Cobalt Complexes and Equilibrium ... 257
- E9.11 Dissolving Aluminum Hydroxide by Complex Reactions ... 258
- E9.12 Dissolving Silver Halogenides by Complex Reactions ... 258

10.5 Experiments on Energy ... 279
- E10.1 Masses and Exothermic Reactions ... 279
- E10.2 Energy Transfer During Exothermic Reaction ... 280
- E10.3 Energy Transfer During Endothermic Reaction ... 280
- E10.4 Temperatures of Ice–Water Mixtures ... 281
- E10.5 Specific Heats of Water and Glycol ... 281
- E10.6 Efficiency Level ... 282
- E10.7 Quantitative Combustion of Pure Carbon ... 283
- E10.8 Quantitative Oxidation of Butane ... 283
- E10.9 Explosion of Gasoline–Oxygen Mixtures ... 284
- E10.10 Molecular Models for Hydrocarbon–Oxygen Reactions ... 285

Epilogue

Misconceptions: They are hard; they have no rules; they are unstructured! The caption in one of Gary Larson's comics seems to best summarize this: "*Shhhh, Zog!... Here comes one now!*"

The work in this book represents a thorough and significant advance in the study of student misconceptions in chemistry. The research reported here, conducted in Germany, the United States, and Ethiopia, explored several common misconceptions about chemistry, and examined diverse ideas of conceptual change that describe how to correct or prevent such misconceptions.

This research adds to existing studies in two major ways: (1) it extends the diagnostic study of students' misconceptions, and (2) it uses a research methodology that allows for the testing of new approaches and strategies to avoid and cure this problem. In this direction one has to differentiate preconcepts and school-made misconceptions: the first ones are brought by young students from observing every-day life, the second ones are developed by inappropriate teaching or by difficult chemistry contents.

Empirical studies of the last decades show that in many countries the same misconceptions are found year after year and nothing changes to improve the situation. So the authors suggest several meaningful ways to prevent and deal with students' chemistry misconceptions. The first reported way is about students initially performing key experiments, followed by discussion of their ideas and understanding, and finally teaching the chemistry concept. The second approach starts by teaching the scientific concept, recounting students' individual misconceptions, finally comparing and defending the new attained chemistry concept by constructively criticizing their prior knowledge. Both ways involve the use of structural models and model drawings which help in the development of students' mental models to facilitate their understanding of the 3-D arrangement of atoms and ions – the base of understanding chemistry. In addition, students will improve their spatial ability with the use of structural models.

We urge everyone to apply the ideas in this book in his or her classroom presentations. Even with limited time and materials, some of the smaller, more simple changes can be accomplished. The authors hope that this book will stimulate serious discussions at all educational institutions to cure preconcepts and successfully to prevent school-made misconceptions. We expect that these fruitful debates and studies will also take into account the need for new or reallocated methods to implement and support improved presentations of chemical material. We welcome comments about this work from students, teachers, and college-level faculty members.

In conclusion, we believe that the research and practical contributions of our studies will inspire future research in this area. Indeed, the design and the results of these studies will provide useful background for replication and expansion to other areas and topics in school and college STEM (Science, Technology, Engineering and Mathematics) courses. College STEM departments could also invite colleagues from schools or colleges of education to focus on issues of teacher preparation and professional development. This will also help improve the scientific literacy of *all*.

A final thought: We would like to quote Albert Einstein, *"You do not really understand something unless you can explain it to your grandmother"*. This is not just a cute and interesting statement, but it is also true. We hope that, with the help of to the material presented in this book, more grandmothers will receive clear and truthful explanations and fewer chemistry misconceptions.